Industrial Utilization of Surfactants

Industrial Utilization of Surfactants

Principles and Practice

Milton J. Rosen
Director, Surfactant Research Institute
Brooklyn College
City University of New York

Manilal Dahanayake
Senior Manager and Scientist Fellow
Rhodia, Inc.
Cranbury, N.J.

AOCS PRESS
Champaign, Illinois

AOCS Mission Statement

To be a forum for the exchange of ideas, information, and experience among those with a professional interest in the science and technology of fats, oils, and related substances in ways that promote personal excellence and provide high standards of quality.

The paper used in this book is acid-free and falls within the guidelines established to ensure permanence and durability.

Library of Congress Cataloging-in-Publication Data

Rosen, Milton, J.
Industrial utilization of surfactants : principles and practice / M.J. Rosen, M. Dahanayake.
p. cm.
Includes bibliographical references and index.
ISBN 1-893997-11-1
1. Surface active agents—Industrial applications. I. Dahanayake, M. (Manilal) II. Title.

TP994.R66 2000
668'.1—dc21 00-064287
CIP

Printed in the United States of America with vegetable oil-based inks.

00 5 4 3 2 1

Preface

There has been a plethora of books on surfactants, surfactant-containing products, and surfactant-related processes in recent years. In spite of this, the selections of surfactants for a particular application still is too often done on a trial-and-error basis. This is partly because the basic principles involved in surfactant utilization are often obscured by the mathematics required to apply these principles rigorously and partly because most real-life applications are too far removed from the simplified models of these applications to which these principles can be applied rigorously.

What the average technologist, who wishes to select in rational fashion a surfactant for use in some process or product, needs is information on the relationship between the *chemical structure of the surfactant and its performance* in that application. A knowledge of the mathematical equations is fundamental to an understanding of the relevant interfacial phenomena, but unless these equations lead to an understanding of how the chemical structure of the surfactant affects these phenomena, these equations have limited value to the technologist who must select a chemical, not an equation, from his laboratory shelf. Such chemical structure–performance property relationship are too often missing from most of the books published on surfactants.

The aims of this book are as follows:

- To provide this chemical structure–performance property information in the simplest possible terms.
- To provide the principles underlying the relationships of chemical structure to performance properties, so that the reader can understand why particular surfactant structures have certain performance properties.
- To give some examples of the application of these relationships to real-life uses of surfactants.

Those who would like to explore chemical structure–fundamental property relationships in somewhat greater depth might consult Rosen, M.J., *Surfactants and Interfacial Phenomena*, 2nd edition, John Wiley, New York, 1989.

We have specifically omitted application areas, such as laundry detergents and cosmetics, in which considerable work has been published and instead have covered areas where much less information is available.

Milton J. Rosen
Manilal Dahanayake

Contents

Industrial Utilization of Surfactants

CHAPTER 1

General Principles

Which Surfactant Should I Use?

A recent compilation of commercially available surfactants by trade name lists several thousand materials made in the United States alone (1). Undoubtedly, most of these are very similar products made by several different manufacturers. Even so, hundreds of different chemical structures can be found among these products. How does one decide which one to use for a particular application? Very often, the trial-and-error method is used: surfactants are picked at random from the shelf in the laboratory, with the hope that one will do the job. At this stage in our knowledge of surfactant science, however, we should be able to do better than that. Although it is still impossible to pinpoint the *exact* surfactant chemical structure that is best for a particular process or product, selection in a rational, scientific manner of the proper structural types for a particular use is possible. The objective of this volume is to show how this selection can be done by describing the principles involved in surfactant utilization and then showing some applications of these principles to actual commercial use.

At the molecular level, a surfactant is an organic compound (Fig. 1.1) that contains at least one lyophilic ("solvent-loving") and one lyophobic ("solvent-fearing") group in the molecule. If the solvent in which the surfactant is to be used is water or an aqueous solution, the respective terms are hydrophilic and hydrophobic. When the surfactant is dissolved in a solvent, the presence of the lyophobic group distorts the normal structure of the solvent, increasing the free energy of the system, and the surfactant then spontaneously orients itself in some manner that will minimize contact between the solvent and the lyophobic group, thereby decreasing the free energy of the system. Thus in aqueous media, the surfactant molecules may migrate to the interfaces of the system and orient themselves there in such fashion as to minimize, as much as possible, contact between the water and their hydrophobic groups. This process is known as **adsorption**, and it changes the properties of the interfaces. This will be discussed in Chapter 2.

Another method of minimizing contact between the lyophobic group of the surfactant molecule and the solvent is by aggregating the

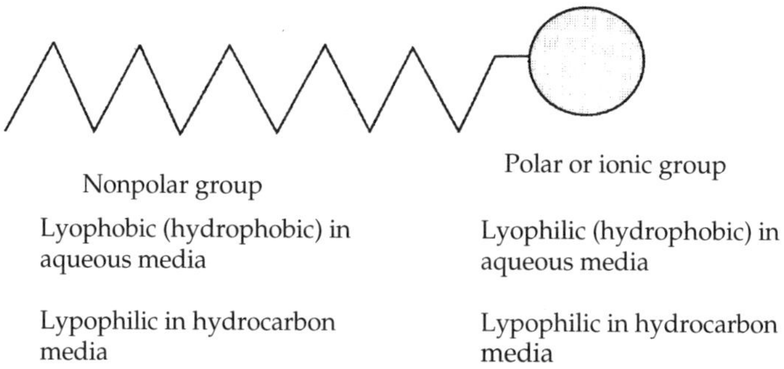

Figure 1.1. Generalized surfactant structure.

surfactant molecules into structures such as spheres, cylinders, or sheets, where the lyophobic groups are in the interior of the aggregate structure and the lyophilic groups are at the exterior, facing the solvent. Such a structure is called a **micelle** (Fig. 1.2). The process is called micellization and changes the properties of the solution phase. It will be discussed in Chapter 3.

The lyophilic group of the surfactant serves to keep it in solution. To retain those properties that are characteristic of surfactants—their **surface activity**—the surfactant must remain soluble in the solvent. For example, if the surfactant becomes insoluble in the solvent, it loses much if not all of its ability to cause the solution to wet surfaces or produce foam.

Surfactants, then, because of the presence of the lyophobic group in the molecule, produce two separate changes in the system: (i) they change the properties of the interfaces, and (ii) they change the properties of the solution phase.

Hydrophobic groups present in commercially available surfactants (Table 1.1) are usually either: (i) a hydrocarbon residue, (ii) a perfluoro-

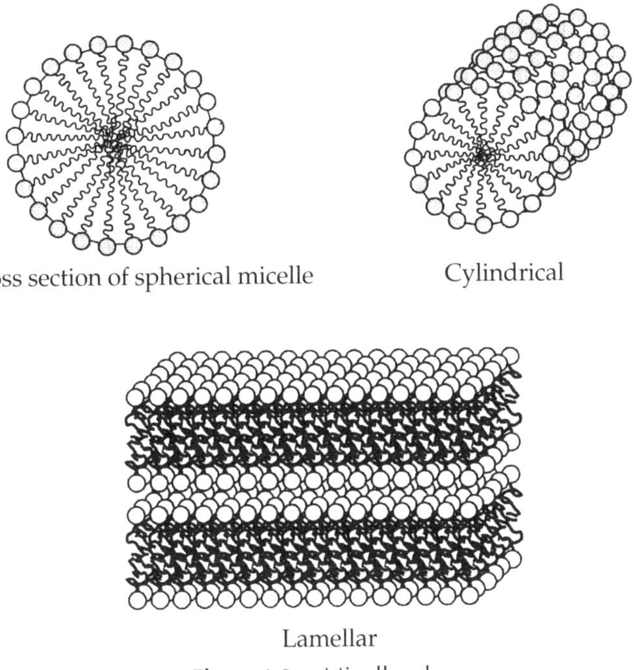

Figure 1.2. Micellar shapes.

TABLE 1.1
Some Hydrophobic Groups in Commercially Available Surfactants

Linear, saturated alkyl (n-dodecyl)	$CH_3(CH_2)_{10}CH_2$–
Branched, saturated alkyl (2-ethylhexyl)	$CH_3(CH_2)_3CH\ CH_2$– (with CH_2CH_3 branch on CH)
Linear, unsaturated alkyl (oleyl)	cis–$CH_3(CH_2)_7CH = CH(CH_2)_7CH_2$–
Alkylbenzene (linear dodecylbenzene)	$CH_3(CH_2)_{11}C_6H_4$–
Alkyldiphenyl ether	$C_6H_5O\ C_6H_4(R)$–
Polyoxypropylene	$-[OCH(CH_3)CH_2]_x^-$
Polyoxybutylene	$-[OCH(C_2H_5)CH_2]_x^-$
Polysiloxane	$(CH_3)_3Si[OSi\ (CH_3)]_xOSi(CH_3)_3$
Perfluoroalkyl	$CF_3(CF_2)_xCF_2$–
Lignin	Complex polymeric phenol

hydrocarbon residue, (iii) a siloxane residue, or (iv) a polyoxypropylene or polyoxybutylene residue. The hydrocarbon residue may be straight alkyl chain, branched alkyl chain, saturated, unsaturated, partly cyclic, or aromatic. The perfluorohydrocarbon residue may be straight chain or branched chain, completely perfluorinated, or attached to a hydrocarbon residue. Siloxane residues are often attached *via* short alkyl chains to the lyophilic group.

The hydrophilic group (Table 1.2) attached to the hydrophobic group may be: (i) negatively charged, in which case the surfactant is anionic, e.g., $C_{12}H_{25}SO_4^-$; (ii) positively charged, in which case the surfactant is cationic, e.g., $C_{10}H_{33}N(CH_3)_3^+$; (iii) both positively and negatively charged, i.e., a zwitterion, often called an ampholyte (i.e., an amphoteric surfactant), e.g., $C_{14}H_{29}N^+(CH_3)_2CH_2COO^-$; or (iv) without formal charge, i.e., a nonionic surfactant, e.g., $C_{12}H_{25}(OC_2H_4)_8OH$.

Whether a group is lyophilic or lyophobic depends on the nature of the solvent in which the surfactant is dissolved for use. Thus in water solution, an alkyl chain is lyophobic (hydrophobic) and an ionic group is lyophilic (hydrophilic). On the other hand, in hydrocarbon solution, the alkyl chain is lyophilic (compatible with the hydrocarbon solvent), whereas the ionic group is lyophobic (causing distortion of the hydrocarbon structure). The nature of the solvent used must therefore be considered in determining which group is lyophilic and which is lyophobic.

TABLE 1.2
Some Hydrophilic Groups in Commercially Available Surfactants

Anionics	
Sulfate	$-OSO_2O^-$
Sulfonate	$-SO_2O-$
Phosphated ethoxylates	$-[(OC_2H_4)_x]_2\ P(O)O^-$ $-[(OC_2H_4)_x]\ P(O)(O^-)_2$
Carboxylate	$-COO^-$
Cationics	
Ammonium, primary	$-NH_3^+$
Ammonium, secondary	$-NH_2^+$ (with one additional bond above N)
Ammonium, tertiary	$-NH^+$ (with additional bonds above and below N)
Ammonium, quaternary	$-N^+-$ (with additional bonds above and below N)
Pyridinium	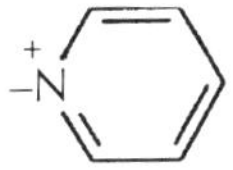
Nonionics	
Polyoxyethylene (ethoxylate)	$-(OCH_2CH_2)_xOH$
Monoglyceride	$-OCH_2\ CHOH\ CH_2OH$
Diglyceride	$-OCH_2\ CH(O-)\ CH_2OH$ $-OCH_2\ CHOH\ CH_2O-$
Acetylenic glycol	$-C(OH)-C \equiv C-C(OH)-$ (each C with an additional bond above)
Pyrrolidinone	

(*continued*)

TABLE 1.2
(Continued)

Nonionics (continued)	
Monethanolamide	$-NH\ CH_2\ CH_2\ OH$
Diethanolamide	$-N(CH_2CH_2OH)_2$
Polyglycoside	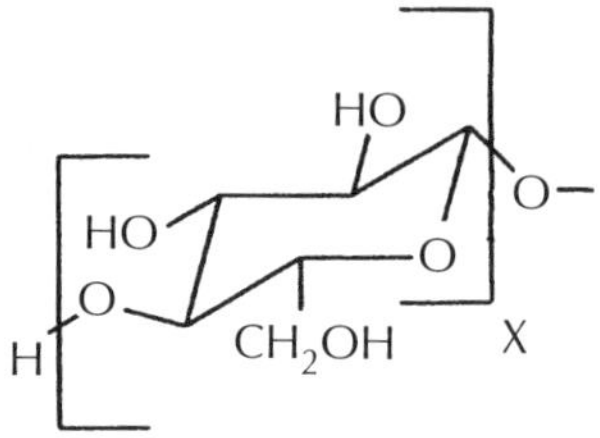
Sorbide	$CH_2\ (CHOH)_3\ CH\ CH_2\ O-$ (with –O– bridge linking first CH_2 and CH)
Zwitterionics[a]	
Aminocarboxylates	$-\overset{+}{N}H_2\ (CH_2)_x\ COO^-$, $-N^+H(CH_2)_x\ COO^-$
Betaine	$-N^+(CH_2)_x\ COO^-$
Sulfobetaine	$-N^+(CH_2)_x\ CH_2SO_3^-$
Amine oxide	$-N^+-O^-$

[a]Often called ampholytic or amphoteric (incorrectly, in the case of betaines and sulfobetaines).

Which Surfactant Is "Best"?

What Do We Mean by "Best"? Before trying to answer that question, we must first decide what we mean by "best." Do we mean the level of performance the surfactant can make available (its effectiveness), do we mean how much surfactant is needed to attain the desired level of performance (its efficiency), or do we mean how fast the surfactant can make available the desired level of performance (its speed of action)?

These all depend on different properties of the surfactant. Thus the efficiency and effectiveness of the surfactant, as we shall see, depend on its equilibrium properties, whereas its speed of action often depends on its dynamic (nonequilibrium) properties.

Other considerations also determine which surfactant is best for a particular use, some physico-chemical, some not.

Chemical Stability. The chemical stability of the surfactant in the system in which it is to be used is generally an important consideration. In some cases, instability is desired. For example, in formulating coatings with a surfactant, it is often desirable for the surfactant to lose its hydrophilic properties during drying of the film, in order that the final film not be sensitive to moisture in the air or water. On the other hand, in formulating a cosmetic emulsion that may be used over a time of considerable length, it is necessary that the surfactant(s) used not hydrolyze or undergo any other decomposition that may result in demulsification of any of the ingredients.

Ester linkages, –C(O) –OR, found in glycerol, sorbitol, polyglycol, polysaccharide, and other polyol esters, hydrolyze in both aqueous alkaline and acidic media, whereas organic sulfate linkages, $-COSO_3^-$, found in sulfated alcohols and sulfated ethoxylated alcohols, are stable in alkaline media but hydrolyze in aqueous acidic media. The acetal

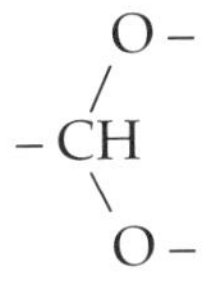

linkage, found in alkyl polyglycosides, is also stable in aqueous alkaline media, but hydrolyzes in aqueous acidic media:

The anionic hydrophilic carboxylate group, $-COO^-$, is converted to the less water-soluble nonionic carboxylic acid group, –C(O)–OH, in acid media, whereas the cationic (RNH_3^+, R_2NH_2, and R_3NH^+) ammonium salts are transformed to their corresponding less water-soluble nonionic free amines in basic media. Surfactants containing chemical groups (e.g., 1 and 2° hydroxyl, sulfide, amino, and benzylic hydrogen groups) that are sensitive to oxidizing agents cannot be used with oxidizing ingredients such as hypochlorite bleach (unless these groups are sterically blocked from reacting).

Environmental Impact. A major concern at present is the effect of chemicals on the environment, and this extends also to surfactants. Consequently, in deciding which surfactant to use for a particular purpose, one of the considerations may be its environmental impact. Here, the relevant properties are: (i) its biodegradability, and (ii) its toxicity to beneficent organisms it may encounter, both in its use and in its disposal. Generally, the greater the rate of biodegradation to innocuous products, the less important is the toxicity of the undegraded surfactant and its degradation products. The volume of surfactant used may also determine the importance of biodegradation and toxicity. Surfactants used in small volume for specialized purposes can be expected to have less environmental impact than those used in large amounts.

Biodegradability (2). Surfactants containing alkyl groups that are branched are more resistant to biodegradation than those containing straight alkyl chains, particularly if the branching is adjacent to the terminal methyl group of the chain. A general principle appears to be that increased distance between the hydrophilic group and the far end of the hydrophobic group increases the speed of biodegradation. The speed of degradation decreases with an increase in the number of methyl groups on the alkyl chain and with the number of alkyl chains in the molecule. The ease of biodegradability also decreases with an increase in the number of oxyethylene groups in ethoxylated surfactants. Secondary alcohol ethoxylates degrade more slowly than primary alcohol ethoxylates, even when the former alcohol is linear. The replacement of oxyethylene groups by oxypropylene or oxybutylene groups also decreases biodegradability.

Toxicity. The toxicity of surfactants depends on both their tendency to adsorb onto organisms and the ease with which the surfactant molecule can penetrate cell membranes (3): tendency to adsorb onto organisms appears to increase with tendency of the surfactant to adsorb onto related interfaces (see Chapter 2, Adsorption onto Insoluble Solids and Liquids from Nonaqueous Solutions of Surfactants), whereas ease of penetration appears to decrease with an increase in the area occupied by the surfactant at the interface (see Chapter 2, Adsorption onto Insoluble Solids or Liquids from Aqueous Solutions of Surfactants). Another property related to its environmental impact is the aqueous solubility of the surfactant in the presence of other water-soluble or water-dispersible substances with which it may come in contact, since often the effect of a surfactant on other organisms depends on its concentration in aqueous media that contact the organisms. For example, hardness (Ca^{2+}, Mg^{2+}) in the water may precipitate an anionic surfactant from its solution in a waste water stream and consequently reduce its environmental impact.

Skin Irritation. In products that may come in contact with the skin, skin irritation by the surfactant, *in the formulation in which it contacts the skin*, is a major factor in deciding which surfactant to use. Protein denaturation, as a result of adsorption of the surfactant onto charged sites on the skin, is believed to be one of the major causes of skin irritation. Here, as in the preceding discussion of toxicity, the tendency of the surfactant to adsorb onto the skin and the ease with which it can penetrate the cell membranes appear to determine skin irritation. Studies of protein denaturation by surfactants indicate that both anionic and cationic surfactants may produce considerable denaturation of proteins. In general, the order of denaturation is: anionics, cationics > amphoterics, amine oxides > ethoxylated nonionics (4–6). For anionic surfactants with C_{12} alkyl chains, the order is: alkylbenzenesulfonate > alcohol sulfate > alpha olefin sulfonate > ethoxylated alcohol sulfate. For the series $C_{12}H_{25}(OC_2H_4)_xSO_4Na$, no denaturation occurs when $x = 6$ or 8 (6).

The addition of positively charged organic material (surface-active or otherwise) that interacts with anionic surfactants can decrease the skin irritation of the latter. This is probably because the positively charged material interacts with the anionic surfactant and decreases its tendency to adsorb onto positively charged sites on the skin. The greater the positive charge of the additive, the greater its protective action on the skin. Thus, positively charged protein hydrolysates, when added to anionic detergents, protect the skin from irritation (7). The addition also of cationic surfactants or long-chain amine oxides to anionic surfactants decreases the irritation of the skin.

Finally, economic considerations, such as cost of the surfactant and the final product or process in which it appears, are of major importance in determining which surfactant to use. These are outside the scope of this book, however.

What Do I Want the Surfactant(s) to Do?

As mentioned in the preceding discussion, surfactants have two sets of properties: (i) they adsorb at interfaces and, as a result of this adsorption, change the properties of those interfaces; and (ii) they aggregate in the solvent in which they are dissolved and, as a result of this aggregation, change the properties of the solution phase. In order to decide which surfactant(s) to use, one must first decide whether, to achieve the performance desired, the properties of the interface(s) must be changed, or the properties of the solution phase must be changed, or both. Since performance is generally observable as a macroscopic property, such as foaming, rather than at the molecular level, e.g., as formation of a monomolecular film at an interface, it is important to understand the relationship of these macroscopic "performance properties" to the molecular level changes in the properties of the interface(s) or the solution phase. These will be discussed in the next two chapters.

Surfactants are used because their presence in the product or process results in such macroscopic phenomena (performance properties) as: (i) wetting or waterproofing, (ii) foaming or defoaming, (iii) emulsification or demulsification, (iv) dispersion or flocculation of solids in liquids, (v) solubilization of solvent-insoluble material in a solvent, or (vi) viscosity increase or decrease of the solution phase. Cleaning of a substrate (detergency), the "performance property" for which surfactants are most commonly used, involves several of the above phenomena.

Performance Properties That Involve Changing the Properties of the Interface(s)

Wetting and Waterproofing. Wetting and waterproofing are "performance phenomena" that depend on changes produced by surfactants in the nature of interfaces. In wetting, a liquid spreads over a substrate (liquid or solid) and displaces the phase that was originally in contact with the substrate, replacing it by a layer of the spreading liquid, which now has new interfaces with both the substrate and the phase originally in contact with it.

The most common example is the wetting of a substrate by water or an aqueous solution (Fig. 1.2). Since water does not spread on hydrophobic ("water-hating") substrates, such as waxy or oily surfaces, surfactants of the proper structure ("wetting agents") are added to the water to cause it to spread. The water spreads as a result of changes in the properties of the interfaces of the system resulting from adsorption of the surfactant onto them.

Waterproofing is similar to wetting in that it involves changing the nature of an interface of the system by adsorbing a surfactant onto it. It differs from the wetting situation just mentioned in that the substrate/air interface to be waterproofed is rendered more hydrophobic than it originally was by adsorption of an added surfactant onto it; this process makes wetting by water more difficult.

Foaming and Defoaming. Foaming and defoaming depend on the surfactants producing changes in the properties of the gas/solution interface. Foam is produced when a gas is introduced into a solution whose surface film has viscoelastic properties. Pure liquids do not foam. The adsorption of a surfactant with the proper structure (a "foaming agent") at the gas/solution interface imparts viscoelastic properties to the interfacial film, and the solution foams. In defoaming, the added surfactant (a "defoaming agent") eliminates or diminishes the existing viscoelastic properties of the gas/solution interfacial film, either by neutralizing or eliminating the surfactant originally there or by replacing the original film by one having poor or no viscoelastic properties.

Emulsification and Demulsification. An emulsion is an opaque dispersion of one liquid (the "discontinuous phase") in a second, immiscible liquid (the "continuous phase").* The emulsion is stabilized by a surfactant film ("emulsifying agent") at the interface between the two liquids that produces electrical or steric barriers to coalescence of the droplets of the dispersed liquid phase. Demulsification of an existing emulsion occurs when these electrical or steric barriers are reduced or eliminated, resulting in a "breaking" of the emulsion. Demulsification can be produced by physical or chemical means. Surfactants whose addition to an existing emulsion cause it to break are called "demulsifying agents." Their structures vary greatly, depending on the nature of the two liquids in the emulsion and the emulsifying agent(s) present.

Dispersion and Flocculation of Solids in Liquids. As in emulsions, the dispersion of particles of a solid in a liquid in which it is insoluble is stabilized by a surfactant film ("dispersing agent") at the interface between the two phases that produces electrical or steric barriers to the aggregation of the dispersed solid particles. The reduction or elimination of these electrical or steric barriers causes flocculation.

---

*Actually, a "macroemulsion," to distinguish it from a "microemulsion," which will be discussed in Performance Properties That Involve Changing the Properties of the Solution Phase.

Adhesion Promotion. Adhesion between two immiscible phases depends on the strength of the interaction between the two different molecules facing each other across the interface between them. The stronger the interaction between the two, the greater the adhesion between the two phases. Interactions across an interface are strong between hydrophobic groups of similar nature (hydrocarbon-hydrocarbon, siloxane-siloxane, perfluorohydrocarbon-perfluorohydrocarbon) and between oppositely charged dipoles or ions. Since the nature of an interface can be changed by the adsorption onto it of a surfactant, this can be used to increase (or decrease) the adhesion between a substrate and a second phase. Thus, the adhesion of a hydrophobic phase to a substrate that has a hydrophobic surface can be increased by the adsorption onto the substrate of a surfactant layer that will be oriented with its hydrophilic group toward the hydrophilic groups of the substrate and its hydrophobic groups oriented away from it. With this adsorbed surfactant layer, the surface of the substrate will now be hydrophobic and a hydrophobic phase of similar nature will now adhere much more strongly to it than to the original hydrophilic surface.

All of the above phenomena will be discussed in greater detail in Chapter 2.

Performance Properties That Involve Changing the Properties of the Solution Phase

Solubilization of Solvent-Insoluble Material. Most commonly, this is desired when the solvent is water, i.e., when we want to solubilize a water-insoluble material in water or some aqueous solution. This has come to be of great importance in recent years because of restrictions on the use of many organic solvents formerly used to dissolve water-insoluble material, in response to the possible negative environmental impact of those organic solvents. However, important situations still exist in which it is desired to solubilize solvent-insoluble material in nonaqueous solvents. Particularly noteworthy is the solubilization of water into fuels, especially airplane fuel, to prevent the formation of ice crystals in fuel lines at temperatures below the freezing point of water.

Solubilization of solvent-insoluble material depends on the presence of surfactant micelles in the solvent phase. Water-insoluble material is solubilized in aqueous media in the hydrophobic interior of the surfactant micelles, either in the outer or inner portions, depending on the polarity of the solubilized material. In either case, the size of the micelles, including solubilized material, is so small that the solution is optically clear, or almost so. Most importantly, solubilization produces a thermodynamically stable system that is consequently indefinitely stable. This is in contrast to the (macro)emulsification of water-insoluble liquids or the dispersion of particles of water-insoluble solid material discussed previously, which produce thermodynamically unstable systems that are not optically clear and which eventually must "break" or separate into two phases. In nonpolar (e.g., hydrocarbon) solvents, micelles can be formed with the polar or ionic hydrophilic groups of the surfactant in the interior portion. Water-soluble, hydrocarbon-insoluble material can be solubilized in this interior portion to form an optically clear, thermodynamically stable solution.

Under certain conditions (to be discussed in Chapter 3), solutions containing micelles can solubilize relatively large amounts of both water and hydrocarbons in approximately equal proportions. These systems are called microemulsions, to distinguish them from (macro)emulsions. In contrast to the latter, which are opaque and thermodynamically unstable, microemulsions are transparent and thermodynamically stable.

Hydrotropy. This is a property that some surfactants or surfactant-like molecules have of increasing the solubility of various solutes in the solution.

Viscosity Increase. This phenomenon is also a function of the micelles in the system. The viscosity of the solution phase increases with an increase in the volume fraction of the dissolved material (solute) in the solution, but is even more dependent on the structure of the surfactant micelles formed there and the way in which these micelles pack together.

The range of these properties will be discussed in Chapter 3.

Performance Properties That Involve Changing the Properties of Both the Interface(s) and the Solution Phase

In some performance properties, changes in both interfacial properties and solution properties are required to optimize performance. Thus foaming, discussed earlier under Performance Properties That Involve Changing the Properties of the Interface(s), is often increased by increasing the viscosity of the liquid in which the air bubbles are dispersed. Emulsions, also mentioned in that section, may be stabilized by liquid crystals in the "continuous phase" or by an increase in its viscosity, due to the presence in it of surfactant(s) of the proper structure. Detergency, the complex performance property for which surfactants are most commonly used, may involve wetting, emulsification, solubilization, and the dispersion of solid particles by the detergent solutions.

References

1. *McCutcheon's Emulsifiers and Detergents 1999*, McCutchcon Division, MC Publishing, Glen Rock, New Jersey, 1999.
2. Swisher, R.D., *Surfactant Biodegradability*, Dekker, New York, 1987, p. 424 ff.
3. Rosen, M.J., L. Fei, Y.-P. Zhu, and S.W. Morrall, *J. Surfactants Detergents 2*:343 (1999).
4. Miyazawa, K., M. Ogawa, and T. Mitsui, *Int. Cosmet. J. Sci. 6*:33 (1984).
5. Rhein, L.D., C.R. Robbins, K. Fernec, and R. Cantore, *J. Soc. Cosmet. Chem. 37*:125 (1986).
6. Ohbu, K., N. Jona, N. Miyajima, and M. Fukuda, *Proceedings of the 1st World Surfactants Congress*, Munich, 1984, Vol 3, p. 317.
7. Tavss, E.A., E. Eigen, V. Temnikov, and A.M. Kligman, *J. Amer. Oil Chem. Soc. 63*:574 (1986).

CHAPTER 2

How the Adsorption of Surfactants Changes the Properties of Interfaces and Related Performance Properties

From the discussion in Chapter 1, we should now know what we want the surfactant to do in a particular situation: change the properties of interface(s), or change the properties of the solution, or both. The next question, then, is: how does it do that? In the situations in which we wish the surfactant to change the properties of the interface(s) of the system, this is accomplished by adsorption of the surfactant at the relevant interface(s).

Let us then examine when and how surfactants can adsorb at interfaces and what happens as a result of this adsorption.

First, we must know some fundamental concepts about interfaces:

1. An **interface** is the contact where two phases meet. A **surface** is an interface where one of the two phases in contact is a gas, usually air.
2. Work is required to produce an interface or a surface between two phases. The work required to produce a unit area of interface (or surface) is called the **interfacial** (or **surface**) **tension**. The minimum work (W_{min}) to increase an interfacial (or surface) area (ΔA) is the product of interfacial (or surface) tension, γ_I, and the change in area:

$$W_{min} = \gamma_I \times \Delta A \qquad [2.1]$$

   When an interfacial (or surface) area is reduced, the work originally required to produce that interfacial (or surface) area is given back (to the surroundings). Consequently, when some interfacial areas are being increased simultaneously with other interfacial areas being reduced, W_{min} is the sum of the $\gamma_I \Delta A$ products of those interfacial areas that are being increased minus the $\gamma_I \Delta A$ products of those interfacial areas that are being reduced.
3. The interfacial (or surface) tension between two phases is a measure of the dissimilarity in their natures. The more similar their natures, the smaller the interfacial (or surface) tension between them.
4. A surfactant will not adsorb at an interface when this will result in an increase in the interfacial tension (i.e., an increase in the dissimilarity of the groups contacting each other), unless compelled to do so by strong (generally, electrostatic) forces. Consequently, distortion of the structure of the solvent by the lyophobic group is a necessary, but not sufficient, condition for adsorption to occur.

Changes in the Properties of the Surface of a Solution

Aqueous Solutions of Surfactants

Perhaps the simplest case of adsorption is that which occurs at the surface of an aqueous solution. As mentioned in Chapter 1, adsorption at any interface occurs because the presence of the lyophobic group of the surfactant distorts the structure of the solvent, and the system consequently responds to minimize contact between the solvent and the lyophobic group. When a surfactant is dissolved in an aqueous medium, its hydrophobic group distorts the structure of the water (by breaking hydrogen bonds between water molecules and by structuring the water in the vicinity of the hydrophobic group). As a result of this distortion, some of the surfactant molecules are expelled to the interfaces of the system, including the surface, where they orient themselves so as to minimize contact between their hydrophobic groups and water molecules (Fig. 2.1a). The surface of the aqueous solution becomes covered with a single layer of surfactant molecules with their hydrophobic groups oriented predominantly toward the air. Since air molecules are essentially nonpolar in nature, as are the hydrophobic groups, this decrease in dissimilarity between the two groups contacting each other across the surface results in a decrease in the surface tension of the water. Decrease in the surface tension of water is an excellent indication of the adsorption of a surface-active material at the water/air interface. Surface tension values for some commercial surfactants in aqueous media are listed in Table 2.1. Some generalizations regarding the effect of surfactant chemical structure on surface tension values in water are as follows:

1. Minimum surface tension values of ethoxylated compounds with the same hydrophobic group increase with increase in the number

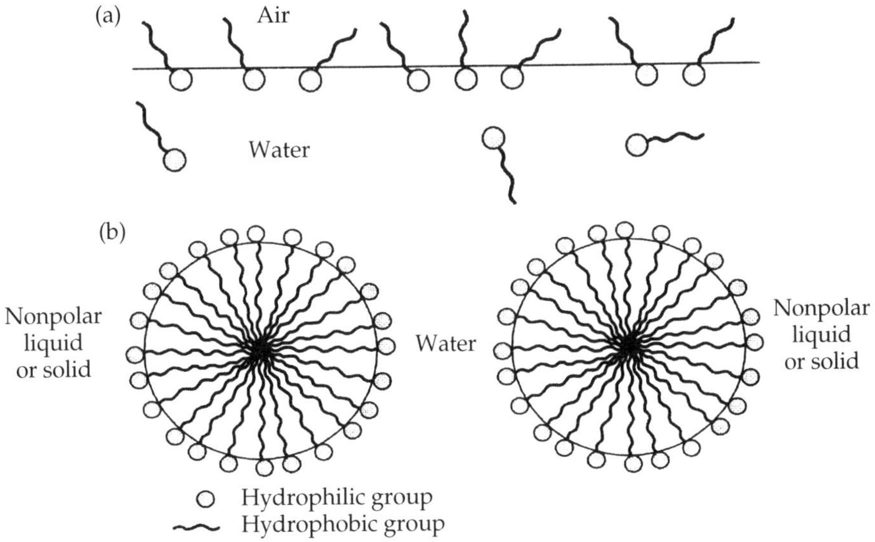

Figure 2.1. (a) Orientation of adsorbed surfactant molecule of air/water interface; (b) orientation of adsorbed surfactant molecules onto nonpolar solid or liquid particles.

TABLE 2.1
Surface Tensions (mN/m) of Some Commercial Surfactants in Aqueous Medium at 22°C

Surfactants	Concentration % (by wt)		
	0.001	0.01	0.1
Nonionics			
(br)*$C_{10}H_{21}(OC_2H_4)_4OH$	54.9	39.2	27.1
(br)*$C_{10}H_{21}(OC_2H_4)_6OH$	45.7	38.7	28.2
$C_{9-11}H_{19-23}(OC_2H_4)_6OH$	53.2	33.4	29.1
$C_{9-11}H_{19-23}(OC_2H_4)_8OH$	54.4	37.1	30.1
$C_{11}H_{23}(OC_2H_4)_5OH$	43.1	26.0	26.0
$C_{11}H_{23}(OC_2H_4)_7OH$	50.2	30.0	28.0
$C_{11}H_{23}(OC_2H_4)_{10}Cl$	48.2	34.1	30.2
$C_{11}H_{23}(OC_2H_4)_{14}Cl$	51.8	37.9	31.0
$C_{12}H_{25}(OC_2H_4)_9OH$	50.9	38.0	32.8
$C_{12}H_{25}(OC_2H_4)_{12}OH$	40.1	35.4	34.2
$C_{12-13}H_{25-27}(OC_2H_4)_7OH$	32.0	28.4	28.2
$C_{12-13}H_{25-29}(OC_2H_4)7OH$	36.0	30.1	28.4
$C_{12-13}H_{25-29}(OC_2H_4)_9OH$	37.4	32.4	29.2
$C_{12-13}H_{25-29}(OC_2H_4)_{12}OH$	39.2	34.2	31.8
$C_{12-13}H_{25-29}(OC_2H_4)_{15}OH$	40.1	35.9	33.9
$C_{12-13}H_{25-29}(OC_2H_4)_{20}OH$	—	—	37.0
$C_{12-13}H_{25-29}(OC_2H_4)_{30}OH$	—	—	37.2
$C_{12-13}H_{25-29}(OC_2H_4)_{40}OH$	—	—	40.1
$C_{12-15}H_{25-31}(OC_2H_4)_7OH$	31.9	30.2	30.4
$C_{12-15}H_{25-31}(OC_2H_4)_9OH$	35.2	31.6	31.4
$C_{14-15}H_{29-31}(OC_2H_4)_7OH$	31.4	29.4	29.7
$C_{14-15}H_{29-31}(OC_2H_4)_{13}OH$	41.0	36.7	34.0
$C_{18}H_{33}(OC_2H_4)_5OH$	33.2	31.8	31.9
$C_{18}H_{35}(OC_2H_4)_{20}OH$	46.4	39.0	37.4
p-t-$C_8H_{17}C_6H_4(OC_2H_4)_5OH$	39.2	29.2	28.2
p-t-$C_8H_{17}C_6H_4(OC_2H_4)_7OH$	44.1	30.1	29.2
p-t-$C_8H_{17}C_6H_4(OC_2H_4)_9OH$	45.9	31.0	30.5
p-t-$C_8H_{17}C_6H_4(OC_2H_4)_{12}OH$	44.4	32.4	31.8
p-t-$C_8H_{17}C_6H_4(OC_2H_4)_{30}OH$	55.4	39.2	35.2
p-t-$C_9H_{19}C_6H_4(OC_2H_4)_6OH$	31.8	28.1	28.0
p-t-$C_9H_{19}C_6H_4(OC_2H_4)_{9-10}OH$	41.2	31.5	31.0
p-t-$C_9H_{19}C_6H_4(OC_2H_4)_{11}OH$	42.5	32.2	30.1
p-t-$C_9H_{19}C_6H_4(OC_2H_4)_{15}OH$	48.4	36.8	36.8
p-t-$C_9H_{19}C_6H_4(OC_2H_4)_{20}OH$	51.1	39.4	38.0
p-t-$C_9H_{19}C_6H_4(OC_2H_4)_{30}OH$	53.4	43.2	39.9
$(C_9H_{19})_2C_6H_3(OC_2H_4)_9OH$	—	29.2	29.2
$(C_9H_{19})_2C_6H_3(OC_2H_4)_{15}OH$	—	30.1	29.4
$(C_9H_{19})_2C_6H_3(OC_2H_4)_{25}OH$	—	33.5	32.8
$C_{8-10}H_{17-21}(C_2H_4O)_4(C_3H_7O)_5H$	48.0	34.2	30.0

(continued)

TABLE 2.1
(Continued)

Surfactants	Concentration % (by wt) 0.001	0.01	0.1
Nonionics (continued)			
$C_{8-10}H_{17-21}(C_2H_4O_6C_3H_7O)_4H$	51.0	38.4	28.4
$C_{6-10}H_{13-21}(OC_3H_7)_4(OC_2H_4)_8(OC_3H_7)_{12}OH$	41.6	34.0	32.7
$C_{17}H_{35}C(O)(OC_2H_4)_5OH$	33.0	32.0	32.0
Castor oil$(OC_2H_4)_{30}OH$	52.1	44.0	41.0
Castor oil$(OC_2H_4)_{40}OH$	58.2	48.5	42.0
$(C_2F_5)_{3-8}CH_2CH_2O(C_2H_4O)_6H$	29.0	24.0	23.0
$C_9H_{19}C(CH_3)_2S(C_2H_4O)_7H$	30.2	29.8	29.8
$[(CH_3)_3SiO]_2Si(CH_3)(CH_2)_3(OC_2H_4)_{7.5}OCH_3$	—	—	20.8
2,4,7,9-Tetramethyl-5-decyn-4,7-diol	—	32.8	32.1
Cationics			
$C_{18}H_{35}N[(C_2H_4O)_{2.5}]_2H$	43.4	33.1	31.3
$C_{18}H_{35}N[(C_2H_4O)_{4.5}]_2H$	44.4	41.2	32.0
Anionics			
$C_8H_{17}SO_4^-Na^+$	69	66.3	53.2
(br)*$C_8H_{17}SO_4^-Na^+$	65.0	61.3	51.2
$C_{12}H_{25}SO_4^-Na^+$ (0.1 *M* NaCl)	33.8	33.1	33.3
$C_{12}H_{21}(OC_2H_4)SO_4^-Na^+$ (0.1 *M* NaCl)	30.2	30.2	30.1
$C_{12}H_{25}(OC_2H_4)_2SO_4^-Na^+$ (0.1 *M* NaCl)	31.2	31.4	31.4
$C_{12}H_{21}(OC_2H_4)_3SO_4^-Na^+$ (0.1 *M* NaCl)	34.2	34.0	34.1
$C_9H_{19}C_6H_4(OC_2H_4)_4SO_4^-Na^+$	54.4	32.4	32.1
$C_9H_{19}C_6H_4(OC_2H_4)_4SO_4^-NH^+_4$	46.4	32.1	31.8
$C_9H_{19}C_6H_4(OC_2H_4)_9SO_4^-NH^+_4$	47.5	37.8	38.3
$C_4H_9CH(C_2H_5)OOCCH_2CH(SO_3^-Na^+)$ $COOCH_2CH(C_2H_5)C_4H_9$	—	43.0	30.4
$C_{12}H_{25}C_6H_3(SO_3^-Na^+)OC_6H_4SO_3^-Na^+$	58.4	37.5	31.4
$C_{11}H_{23}C(O)N(CH_3)CH_2CH_2SO_3^-Na^+$	—	31.8	31.8
$C_{15}H_{31}C(O)N(CH_3)CH_2CH_2\ SO_3^-Na^+$	—	31.4	31.2
$C_{15}H_{31}C(O)N(C_6H_{11})CH_2CH_2\ SO_3^-Na^+$	—	37.0	37.0
$C_{17}H_{33}C(O)N(CH_3)CH_2CH_2\ SO_3^-Na^+$	55.4	30.8	30.2
(Tall oil acyl)$C(O)N(CH_3)CH_2CH_2\ SO_3^-Na^+$	—	40.1	39.4
$(C_2F_5)_{3-8}CH_2CH_2SCH_2CH_2CO_2^-Li^+$	48.4	22.0	18.1
$[C_6H_5(OC_2H_4)_6O]_{1,2}P(O)(OH)_{1,2}$ (pH 3)	69.5	58.2	45.2
$[C_8H_{17}(OC_2H_4)_3O]_{1,2}P(O)(OH)_{1,2}$ (pH 3)	42.5	28.0	26.5
$[C_{12}H_{25}(OC_2H_2)_4O]_{1,2}P(O)(OH)_{1,2}$ (pH 3)	50.9	38.0	32.8
[(br)*$C_{13}H_{27}(OC_2H_4)_4O]_{1,2}P(O)(OH)_{1,2}$ (pH 3)	27.5	27.0	26.5
$[C_{18}H_{35}(OC_2H_4)_4O]_{1,2}P(O)(OH)_{1,2}$ (pH 3)	42.6	38.7	36.5
$[C_9H_{19}C_6H_4(OC_2H_4)_4O]_{1,2}P(O)(OH)_{1,2}$ (pH 3)	31.1	30.2	29.0
$[C_9H_{19}C_6H_4(OC_2H_4)_6O]_{1,2}P(O)(OH)_{1,2}$ (pH 3)	40.0	34.1	32.5

(continued)

TABLE 2.1
(Continued)

Surfactants	Concentration % (by wt) 0.001	0.01	0.1
Anionics (continued)			
$[C_9H_{19}C_6H_4(OC_2H_4)_9O]_{1,2}P(O)(O^-Na^+)_{1,2}$ (pH 7)	48.0	37.3	36.7
$[(C_9H_{19})_2C_6H_3(OC_2H_4)_8O]_{1,2}P(O)(OH)_{1,2}$ (pH 3)	38.0	32.0	29.0
$[(C_9H_{19})_2C_6H_3(OC_2H_4)_{10}O]_{1,2}P(O)(OH)_{1,2}$ (pH 3)	43.5	37.8	34.8
Zwitterionics[a]			
(br)*$C_8H_{17}NH^+(CH_2CH_2COOH)CH_2CH_2COO^-$ (pH 6)	62.0	41.1	27.2
$C_9H_{19}C(O)NH(CH_2)_2N^+H(CH_2CH_2OH)CH_2COO^-$ (pH 6)	63.5	36.1	25.2
$C_{12}H_{25}N^+H_2CH_2CH_2COO^-$ (pH 6)	48.2	25.4	25.5
$C_{12}H_{25}N^+H(CH_2CH_2COO^-)\ CH_2COOH$ (pH 6)	48.0	38.5	38.5
$C_{11}H_{23}C(O)NH(CH_2)_2N^+(CH_2CH_2OH)_2CH_2COO^-$ (pH 6)	28.0	26.5	26.5
$C_{11}H_{23}C(O)NH(CH_2)_2N^+(CH_2CH_2OH)(CH_2COOH)\ CH_2COO^-Na^+$	38.1	29.7	29.7
$C_{11}H_{23}C(O)NH(CH_2)_3N^+(CH_3)_2CH_2COO^-$ (pH 6)	37.2	32.1	32.0
$C_{17}H_{35}C(O)NH(CH_2)_3N^+(CH_3)_2CH_2COO$– (pH 6)	33.7	32.4	32.2
$C_{11}H_{23}CONH(CH_2)_3N^+(CH_3)_2CH_2CH(OH)CH_2SO_3^-$ (pH 6)	45.2	35.4	35.4
$(C_2F_5)_{3-8}CH_2CH(OCOCH_3)CH_2N^+(CH_3)_2CH_2CO_2^-$ (pH 6)	40.0	21.2	19.0
Zwitterionic/anionic mixtures (1:1 molar ratio)			
$C_{11}H_{23}C(O)NH(CH_2)_3N^+H(C_2H_4OH)CH_2C(O)O^-.\ C_{12}H_{25}OC_2H_4SO_4^-Na^+$	26.5	26.4	26.5
$C_{11}H_{23}C(O)NH(CH_2)_2N^+H(C_2H_4OH)_2CH_2C(O)O^-.\ C_{12}H_{25}(OC_2H_4)_2SO_4^-Na^+$	27.9	27.8	27.9
$C_{11}H_{23}C(O)NH(CH_2)_2N^+H(C_2H_4OH)_3CH_2C(O)O$–. $C_{12}H_{25}(OC_2H_4)_3SO_4^-Na^+$	28.4	28.2	28.4

[a]May be anionic at pH >7.
*Branched chain.

of oxyethylene units in the surfactant molecule. However, note that in the $C_{12}H_{25}(OC_2H_4)_x\ SO_4^-Na^+$ series, the lowest minimum surface tension value is when $x = 1$, lower than when $x = 0$ or 2.

2. Surfactants with branched hydrophobic groups show lower surface tension values than their isomeric compounds with straight-chain hydrophobes.

3. Surfactants with perfluoroalkyl chains can yield lower surface tension values (≥15 mN/m) than those with polysiloxane chains, which in turn can yield lower values (≥20 mN/m), than those with alkyl chains (≥25 mN/m).
4. Molecules with closely packed, small hydrophilic head groups give the lowest surface tension values in water.

Reduction of the surface tension is a very important change in the properties of the surface produced there by adsorption of the surfactant, since it underlies many of the performance properties of aqueous surfactant solutions. In many processes of industrial importance, e.g., those involving foaming or wetting, the surface of a liquid must be increased, sometimes by orders of magnitude. Reducing the surface tension of the foaming or wetting solution makes the process easier to accomplish, since it reduces the work required (see Eq. 2.1, p. 16).

Adsorption of a surfactant from aqueous media at the interface with a nonpolar liquid (e.g., a hydrocarbon) is very similar to its adsorption at the surface against air since, as mentioned above, air molecules are essentially nonpolar. The surfactant molecules adsorb at the nonpolar liquid/water interface with their hydrophobic groups oriented predominantly toward the nonpolar liquid phase, and this action results in a decrease in the interfacial tension between them. This reduction in the interfacial tension facilitates any process involving an increase in the interfacial area, such as emulsification or dispersion of the nonpolar liquid, as indicated by Equation 2.1.

Some reductions of interfacial tension against mineral oil by aqueous solutions of commercial surfactants are listed in Table 2.2. Surfactants of borderline solidity in the aqueous media often show low interfacial tensions against mineral oil.

Nonaqueous Solutions of Surfactants

Surfactants can adsorb also at the surface of nonaqueous solutions, but only if their lyophobic groups distort the structure of the solvent. Thus, surfactants with the usual hydrophobic groups can adsorb at the surfaces of highly hydrogen-bonded solvents, such as ethylene glycol solutions, and lower their surface tension against air. On the other hand, surfactants with alkyl chains do not adsorb at the surfaces of many less polar solvents, such as ethyl alcohol, since they do not distort their structure significantly.

In addition, surfactants will not adsorb at the surface of nonaqueous solutions unless by so doing they can reduce the surface tension of the sol-

TABLE 2.2
Interfacial Tensions (mN/m) of Some Aqueous Solutions of Commercial Surfactants Against Mineral Oil at 22°C

	Concentration % (by wt)	
Surfactants	0.01	0.1
Nonionics		
(br)*$C_{10}H_{21}(OC_2H_4)_4OH$	18	8
(br)*$C_{10}H_{21}(OC_2H_4)_6OH$	20	10
$C_{9-11}H_{19-23}(OC_2H_4)_6OH$	16	8
$C_{9-11}H_{19-23}(OC_2H_4)_8OH$	20	11
$C_{11}H_{23}(OC_2H_4)_5OH$	12	6
$C_{11}H_{23}(OC_2H_4)_7OH$	13	8
$C_{12}H_{25}(OC_2H_4)_9OH$	10	7
$C_{12}H_{25}(OC_2H_4)_{12}OH$	12	9
$C_{12-13}H_{25-27}(OC_2H_4)_7OH$	7	5
$C_{12-14}H_{25-29}(OC_2H_4)_7OH$	8	7
$C_{12-14}H_{25-29}(OC_2H_4)_9OH$	10	8
$C_{12-14}H_{25-29}(OC_2H_4)_{12}OH$	12	10
p-t-$C_8H_{17}C_6H_4(OC_2H_4)_5OH$	14	5
p-t-$C_8H_{17}C_6H_4(OC_2H_4)_6OH$	12	6
p-t-$C_8H_{17}C_6H_4(OC_2H_4)_7OH$	11	7
p-t-$C_8H_{17}C_6H_4(OC_2H_4)_9OH$	10	8
p-t-$C_8H_{17}C_6H_4(OC_2H_4)_{12}OH$	13	11
p-t-$C_9H_{17}C_6H_4(OC_2H_4)_5OH$	3	2
p-t-$C_9H_{19}C_6H_4(OC_2H_4)_6OH$	5	3
p-t-$C_9H_{19}C_6H_4(OC_2H_4)_7OH$	5	3
p-t-$C_9H_{19}C_6H_4(OC_2H_4)_{8,5}OH$	6	4
p-t-$C_9H_{19}C_6H_4(OC_2H_4)_9OH$	5	4
p-t-$C_9H_{19}C_6H_4(OC_2H_4)_{10}OH$	6	5
p-t-$C_9H_{19}C_6H_4(OC_2H_4)_{12}OH$	8	7
p-t-$C_9H_{19}C_6H_4(OC_2H_4)_{20}OH$	15	13
p-t-$C_9H_{19}C_6H_4(OC_2H_4)_{30}OH$	17	15
$(C_9H_{19})_2C_6H_3(OCH_2CH_2O)_7OH$	9	7
$(C_9H_{19})_2C_6H_3(OCH_2CH_2O)_9OH$	8	6
$(C_9H_{19})_2C_6H_3(OCH_2CH_2O)_{24}OH$	5	4
$(C_9H_{19})_2C_6H_3(OCH_2CH_2O)_{50}OH$	9	7
Castor oil$(OC_2H_4)_{30}OH$	7	5
Castor oil$(OC_2H_4)_{40}OH$	13	10
Anionics		
$C_{12}H_{25}SO_4^-NH_4^+$	14	9
$C_{12}H_{25}SO_4^-Na^+$	18	11
$C_{12}H_{21}(OC_2H_4)SO_4^-Na^+$	20	11
$C_{12}H_{25}(OC_2H_4)_2SO_4^-Na^+$	24	14

(continued)

TABLE 2.2
(Continued)

Surfactants	Concentration % (by wt) 0.01	0.1
Anionics (continued)		
$C_{12}H_{21}(OC_2H_4)_3SO_4^-Na^+$	25	15
$C_9H_{19}C_6H_4(OC_2H_4)_4SO_4^-Na^+$	29	12
$C_9H_{19}C_6H_4(OC_2H_4)_4SO_4^-Na^+$	28	9
$C_9H_{19}C_6H_4(OC_2H_4)_9SO_4^-NH_4^+$	30	15
$C_4H_9CH(C_2H_5)OOCCH_2CH(SO^-_3Na^+)COOCH_2CH(C_2H_5)C_4H_9$	3	2
$C_{11}H_{23}C(O)OCH_2CH_2SO_3^-Na^+$	—	6
$C_{11-13}H_{23-27}C(O)N(CH_3)CH_2CH_2SO_3^-Na^+$	—	7
$C_{15}H_{31}C(O)N(CH_3)CH_2CH_2SO_3^-Na^+$	—	9
$C_{15}H_{31}C(O)N(C_6H_{11})CH_2CH_2SO_3^-Na^+$	—	5
$C_{17}H_{33}C(O)N(CH_3)CH_2CH_2SO_3^-Na^+$	—	3
$C_{17}H_{35}C(O)N(CH_3)CH_2CH_2SO_3^-Na^+$	—	11
(Tall oil acyl)$C(O)N(CH_3)CH_2CH_2SO_3^-Na^+$	—	16

*Branched chain.

vent against air. Thus, in nonpolar organic solvents such as aliphatic hydrocarbons, surfactants with alkyl chains and polar or ionic groups do not adsorb at the surface, not because they do not distort the structure of the solvent (their polar or ionic groups do), but because in these cases, the ionic or polar groups are the lyophobic groups. If these groups are expelled to the surface, they will *increase* the surface tension of the hydrocarbon phase, since they are now adjacent to nonpolar air molecules, and therefore more dissimilar to them than the original hydrocarbon molecules. Consequently, adsorption at the surface does not occur. Surfactants containing perfluorinated alkyl groups, when the remainder of the molecule is sufficiently lipophilic, can adsorb at the surfaces of hydrocarbon solutions and act as surface tension reducers for them since the perfluoroalkyl group is lyophobic in hydrocarbons and is still similar enough in nature to air molecules to reduce the surface tension against them.

Changes in the Properties of Solid/Liquid and Liquid/Liquid Interfaces

Distortion of the structure of a solvent by a surfactant can cause adsorption of the latter not only onto the surface of the solution but also onto all the other interfaces of the system. Thus, the surfactant can adsorb onto the walls of the container holding the solution and onto any other solids or liquids in contact with the solution.

Aqueous Solutions of Surfactants

If the surfactant-containing solution is aqueous, then the orientation of the surfactant adsorbing onto any nonpolar solid or liquid (second phase) in contact with the aqueous solution will be similar to that at the aqueous solution/air surface described above: the hydrophobic group of the surfactant will be oriented predominantly toward the nonpolar second phase, with the hydrophilic head in the aqueous phase (Fig. 2.1b). Since this orientation (as in the case of orientation toward the nonpolar air) decreases the dissimilarity between the aqueous phase and the nonpolar second phase, the interfacial tension between the two phases is decreased, and it will now be easier than in the absence of the surfactant to increase the area of interface between them. It will therefore be easier to disperse small particles of nonpolar solid or to disperse droplets of nonpolar liquid in the aqueous surfactant phase.

If the surfactant is an ionic type, then adsorption of the surfactant onto the surface of the particles of hydrophobic solid or liquid will produce an electrical charge on the surface of each particle similar in sign to that of the surfactant ion. Since each particle will then have an electrical charge of the same sign, the particles will repel each other and may become dispersed in the aqueous phase. The stability of the dispersion will depend, in part, on the magnitude of the electrical charge on the particles, which produces an electrostatic barrier to their close approach.

If the surfactant is of the nonionic type with a polymeric hydrophilic group (e.g., a polyoxyethylene chain), then adsorption onto the hydrophobic particles will produce a surface layer on each particle with polymeric chains extending into the aqueous phase. This will produce a steric barrier to the close approach of particles to each other and may result in a dispersion, stabilized by this steric barrier, of the hydrophobic particles in the aqueous phase.

On the other hand, if the solid or liquid (second phase) in contact with the aqueous surfactant solution has ionic or highly polar surface groups, then the situation is more complex. If there are ionic or highly polar surface sites on the second phase that can interact attractively with those of the surfactant's hydrophilic group, then the surfactant may adsorb onto those surface sites of the second phase with its hydrophilic group oriented toward them (Fig. 2.2a). Since, in that case, the hydrophobic group of the

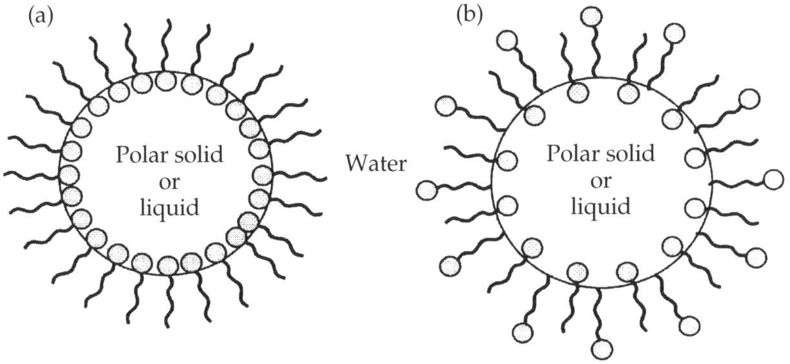

Figure 2.2. Adsorption of surfactant from aqueous solution onto particles of polar solid or liquid that can interact attractively with the hydrophilic head groups of the surfactant. (a) Adsorption of first layer of surfactant. (b) Adsorption of second layer of surfactant.

surfactant will be oriented toward the aqueous phase, with a resulting increase in the aqueous phase/second phase interfacial tension, adsorption with this orientation will occur only if the interaction between the ionic or polar group of the second phase is strong enough to compensate for this increase in interfacial tension. When adsorption with this orientation occurs, the surface of the second phase, because of its layer of strongly adsorbed surfactant, becomes more and more hydrophobic as adsorption continues. Often, at these now hydrophobic areas of the second phase, additional surfactant molecules may adsorb with their hydrophobic groups oriented toward the hydrophobic groups of the previously adsorbed surfactant, either as a second layer on the first adsorbed layer or adjacent to previously adsorbed molecules in a tail-to-tail arrangement (Fig. 2.2b). Molecules adsorbed in this fashion will make the surface of the second phase less hydrophobic. However, since they are adsorbed *via* the relatively weak van der Waals forces between hydrophobic groups, compared to the stronger forces holding the original surfactant layer to polar or ionic surface sites of the second phase, if dilution of the aqueous phase occurs, these more weakly adsorbed surfactant molecules will desorb from the surface of the second phase. This will leave the first layer of strongly adsorbed molecules, with their hydrophobic groups again exposed to the aqueous medium.

On the other hand, if there is little or no attractive interaction between sites on the polar solid or liquid and the hydrophilic head of the surfactant, then no adsorption will occur.

Nonaqueous Solutions of Surfactants

When the surfactant-containing solution is nonaqueous, the extent and orientation of surfactant adsorption depends on the nature of both the solvent in which the surfactant is dissolved and the surface of the second phase. If the two phases are similar in nature (both nonpolar, or both polar), very little surfactant adsorption will occur at the interface, since adsorption of the surface, regardless of its orientation there, will increase the dissimilarity between the two phases of their interface, and consequently the interfacial tension, resulting in the energetically unfavorable increase in the free energy of the system.

If the surfactant is dissolved in a highly polar solvent with multiple hydrogen bonding capacity, such as ethylene glycol or glycerin, adsorption will be similar to that in an aqueous medium (Fig. 2.1), with the hydrophobic group of the adsorbed surfactant oriented away from the highly polar solvent, unless strong interaction exists between the hydrophilic group of the surfactant and the polar or ionic groups on the surface of the second phase. If the surfactant is dissolved in a moderately polar solvent, such as ethanol, adsorption of surfactant at the interface may not occur to any significant extent, because the surfactant, with its mixed hydrophilic-hydrophobic character, may be too similar in nature to the solvent to distort its structure.

If the surfactant is dissolved in a nonpolar solvent, such as a hydrocarbon, then it will usually be adsorbed onto a polar or ionic second phase with its hydrophilic group oriented away from the nonpolar solvent (Fig. 2.3). The only exception is when the surface of the second phase carries an electrical charge of the same sign as that of the

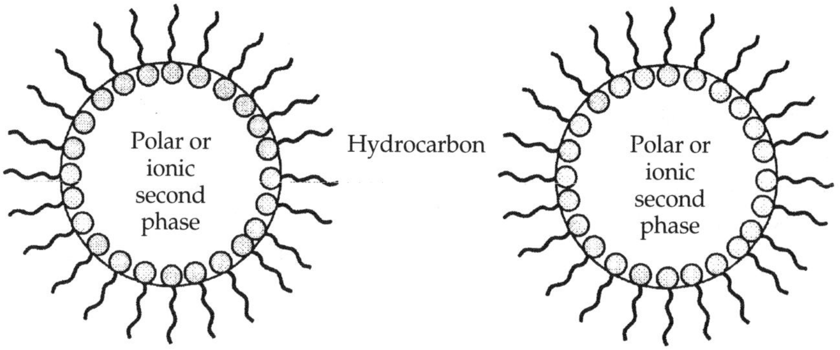

Figure 2.3. Adsorption of surfactant molecules from hydrocarbon solution onto polar or ionic second phase particles of a charge not similar to that of the surfactant.

hydrophilic group of the surfactant. In that case, little or no surfactant will be adsorbed at that interface.

How Quantitative Information on Adsorption at an Interface Is Obtained

The above discussion tells us, in qualitative terms, how surfactant adsorption changes the properties of an interface. However, it does not tell us anything quantitative about the nature of the adsorbed film, e.g., how closely packed the molecules are, or how easily or how rapidly the film forms. These are important to know because they tell us how much, how easily, how quickly, and in what manner the properties of the interface are changed.

Information on the amount of surfactant adsorbed at an interface is generally obtained either directly, by measuring the amount of surfactant that has been lost from the solution phase, or indirectly, by measuring some property of the interface that is known to be related to the amount of surfactant adsorbed there. Generally, the method used depends on the nature of the phase in contact with the phase containing the surfactant.

Adsorption at the Surface of a Surfactant Solution

If we are interested in changes produced by the adsorption of surfactant onto the *surface* of a surfactant solution, then the amount adsorbed can be determined by measuring the equilibrium surface tension (γ) of the solution as a function of the concentration of dissolved surfactant. When the equilibrium surface tension (γ) is plotted (Fig. 2.4) against the log of the molar surfactant concentration ($\log C$), the surface concentration (Γ) can be calculated from the slope of the plot by use of the Gibbs adsorption equation (1[a]):

$$\Gamma = (-\partial\gamma/\partial \log C)_T / 2.303\, nRT \qquad [2.2]$$

where ($\partial\gamma/\partial \log C$) is the slope of the plot and n is the number of species whose concentration at the surface changes when the surfactant concentration changes. Thus, for a nonionic surfactant, $n = 1$; for a surfactant that is a 1:1 electrolyte, $n = 2$ in the absence of added electrolyte. If a swamping amount of electrolyte containing a common counterion is present in the solution containing the 1:1 electrolyte surfactant, then again $n = 1$ (since the amount of counterion does not change when the

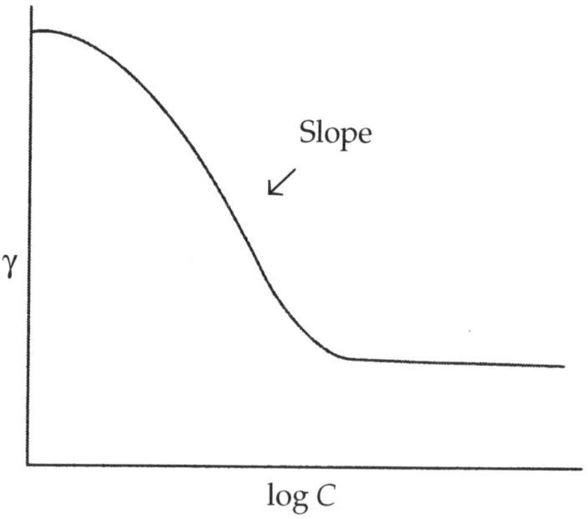

Figure 2.4. Plot of equilibrium surface tension (γ) vs. log of the molar concentration (C) of the surfactant in solution phase.

surfactant concentration changes). When R (the gas constant = 8.31 × 10^7 erg mol^{-1} K^{-1} and the absolute temperature, T, is in K^{-1}, and γ is in dyn/cm (or erg/cm^2), then Γ is in mol/cm^2. When R = 8.31 J mol^{-1} K^{-1} and γ is in mN/m or mJ/m^2, then Γ is in mol/1000m^2.

The area (A) occupied by the surfactant molecule (or ion) at the surface, in Å^2 (= 0.01 nm^2), is then (1[b]):

$$A = 10^{16}/N\Gamma \qquad [2.3]$$

where N is Avogadro's number (6.02×10^{23}), and Γ is in mol/cm^2.

The amount of reduction of the surface tension ($\gamma^{\circ} - \gamma$, where γ° is the surface tension of the solvent in the absence of surfactant) is called the surface pressure, Π.

The area of the surface occupied by the surfactant is a very useful quantity, since it indicates how closely packed the adsorbed molecules are (Chapter 4, Packing at Interfaces). The more closely packed the surfactant film, the more cohesive it will be. This bears directly on performance properties. In some performance properties, it is desirable for the surface film to be cohesive; in others, it is desirable for it to be noncohesive. This will be discussed in the section on Adhesion Promotion.

The plot of γ vs. log C is also an excellent method of determining the **efficiency** (1[c]) and **effectiveness** (1[d]) of the adsorption of the surfactant at the surface. The efficiency of adsorption is a measure, in quantitative terms, of the tendency of the surfactant to adsorb at the surface. It tells us whether the surfactant's tendency to adsorb is strong, moderate, or weak. A good measure of the efficiency of adsorption of a surfactant at an interface is the molar concentration (C) of the surfactant in the solution phase required to produce maximum adsorption of surfactant there. At the aqueous surface/air interface, it has been found empirically that this generally occurs when the surface tension of the solvent has been reduced by 20 dyn/cm (2). The molar concentration of the surfactant in the aqueous phase at this point is called the C_{20} value, and the negative log of the C_{20} value, designated the pC_{20} value, is a good measure of the efficacy of adsorption at the aqueous solution/air interface. A pC_{20} value of 4.0 means that a 1×10^{-4} molar solution of a particular surfactant will reduce the surface tension of the solvent by 20 dyn/cm; a value of 5.0, that a 1×10^{-5} molar solution of a different surfactant will do that. (These concentrations will be close to that needed to "saturate" the surface with surfactant in each case.) We can see from these values that the second surfactant is ten times as efficient as the first. By examining a γ–log C plot of a surfactant, therefore, and seeing what value of log C causes a 20 dyn/cm reduction in the surface tension of the solvent, we can readily determine the efficiency of adsorption of a surfactant at the aqueous solution/air interface.

Some effectiveness of adsorption (Γ_m) and area/molecule (A_m) at surface saturation, and efficiency of adsorption (pC_{20}) values of some commercial surfactants at the aqueous solution/air interface at 22°C are listed in Table 2.3. For ethoxylated nonionics, Γ_m and pC_{20} decrease and A_m increases with increase in the number of oxyethylene units in the molecule. For ionics, Γ_m and pC_{20} increase (and A_m decreases) with increase in the ionic strength of the solution. In all cases, pC_{20} increases with increase in the number of carbons in the hydrophobic group.

Dynamic surface tension data for 0.1% aqueous solutions of some commercial surfactants are listed in Table 2.4. Surfactants with short hydrophobic groups and branched, rather than linear isomeric, ones and with hydrophilic groups that are nonionic or of reduced ionic character, generally yield the lowest surface tension values at short times.

The *effectiveness* of adsorption of a surfactant tells us *how much* surfactant is adsorbed at an interface. We can use as a measure of the effectiveness of adsorption, the maximum concentration that the surfactant can attain at the interface (at "saturation adsorption"). Since the surface concentration (Γ) can be obtained by use of Equation 2.2 from the slope of the γ–log C plot, the maximum concentration (Γ_{max}) of surfactant attainable there can be easily calculated from the maximum slope of that plot, and can be used as a measure of the effectiveness of adsorption. The Γ_{max} value is a major factor determining how much the performance properties are changed by adsorption of surfactant; generally,

TABLE 2.3
Effectiveness of Adsorption (Γ_m) and Area/Molecule (A_m) at Surface Saturation, and Efficiency of Adsorption, (pC_{20}) of Some Commercial Surfactants at the Aqueous Solution/Air Interface (22°C)

Surfactants	Γ_m(mol/cm^2 × 10^{10})	A_m, A^2	pC_{20}
Nonionics			
$C_{9-11}H_{19-23}(OC_2H_4)_6OH$	2.86	58	4.11
$C_{9-11}H_{19-23}(OC_2H_4)_8OH$	2.72	61	3.92
(br)*$C_{10}H_{21}(OC_2H_4)_4OH$	3.46	48	4.11
(br)*$C_{10}H_{21}(OC_2H_4)_6OH$	3.07	54	4.02
$C_{11}H_{23}(OC_2H_4)_5OH$	3.61	46	4.61
$C_{11}H_{23}(OC_2H_4)_7OH$	3.32	50	4.58
$C_{11}H_{23}(OC_2H_4)_{10}Cl$	3.95	42	4.85
(br)*$C_{11}H_{23}(OC_2H_4)_{10}Cl$	3.45	48	4.53
$C_{12}H_{25}(OC_2H_4)_7OH$	3.19	52	5.62
$C_{12}H_{25}(OC_2H_4)_9OH$	2.59	64	5.41
$C_{12-13}H_{25-27}(OC_2H_4)_7OH$	2.96	56	5.32
$C_{12-14}H_{25-29}(OC_2H_4)_7OH$	2.86	58	5.45
$C_{12-14}H_{25-29}(OC_2H_4)_9OH$	2.51	66	5.41
$C_{12-14}H_{25-29}(OC_2H_4)_{12}OH$	2.30	72	5.22
$C_{12-14}H_{25-29}(OC_2H_4)_{15}OH$	1.95	85	4.31
$C_{12-15}H_{25-31}(OC_2H_4)_7OH$	3.07	54	5.52
$C_{12-15}H_{25-31}(OC_2H_4)_9OH$	2.55	65	5.29
$C_{14-15}H_{29-31}(OC_2H_4)_7OH$	3.25	51	5.81
$C_{14-15}H_{29-31}(OC_2H_4)_{13}OH$	2.45	68	5.25
$C_{16}H_{33}(OC_2H_4)_{15}OH$	2.21	75	6.36
$C_{18}H_{35}(OC_2H_4)_5OH$	2.45	68	5.21
$C_{8-10}H_{17-21}(C_2H_4O)_4(C_3H_7O)_5H$	2.86	58	5.24
$C_{8-10}H_{17-21}(C_2H_4O)_6(C_3H_7O)_5H$	2.59	64	5.02
$C_{6-10}H1_{3-21}(OC_3H_7)_4(OC_2H_4)_8(OC_3H_7)_{12}OH$	2.37	70	5.09
p-t-$C_8H_{17}C_6H_4(OC_2H_4)_5OH$	3.19	52	4.82
p-t-$C_8H_{17}C_6H_4(OC_2H_4)_7OH$	3.02	55	4.75
p-t-$C_8H_{17}C_6H_4(OC_2H_4)_9OH$	2.59	64	4.62
p-t-$C_8H_{17}C_6H_4(OC_2H_4)_{12}OH$	2.37	70	4.48
p-t-$C_9H_{17}C_6H_4(OC_2H_4)_6OH$	2.86	58	5.35
p-t-$C_9H_{17}C_6H_4(OC_2H_4)_{9-10}OH$	2.77	60	5.15
p-t-$C_9H_{17}C_6H_4(OC_2H_4)_{11}OH$	2.68	62	4.92
p-t-$C_9H_{19}C_6H_4(OC_2H_4)_{15}OH$	2.37	70	4.42
p-t-$C_9H_{19}C_6H_4(OC_2H_4)_{20}OH$	1.86	89	4.05
p-t-$(C_9H_{19})_2C_6H_3(OC_2H_4)_9OH$	2.86	58	5.35
p-t-$(C_9H_{19})_2C_6H_3(OC_2H_4)_{15}OH$	2.44	68	4.95
$C_{18}H_{35}N[(OC_2H_4)_{2.5}]_2H$	3.45	48	5.41
$C_{18}H_{35}N[(OC_2H_4)_{4.5}]_2H$	2.96	56	5.24
$C_9H_{19}C(CH_3)_2S(C_2H_4O)_7H$	3.45	48	5.41

(*continued*)

TABLE 2.3
(Continued)

Surfactants	Γ_m(mol/ cm^2 × 10^{10})	A_m, A^2	pC_{20}
Anionics			
$C_8H_{17}SO_4^-Na^+$	2.77	60	1.72
(br)*$C_8H_{17}SO_4^-Na^+$	2.45	68	1.62
$C_{12}H_{25}SO_4^-Na^+$	3.81	46	2.56
$C_{12}H_{25}SO_4^-Na^+$ (0.1 *M* NaCl)	4.15	40	3.81
$(C_{12}H_{25}SO_4^-)_2Mg^{2+}$	3.07	54	4.22
$C_{12}H_{21}(OC_2H_4)SO_4^-Na^+$ (0.1 *M* NaCl))	4.02	42	5.07
$C_{12}H_{25}(OC_2H_4)_2SO_4^-Na^+$ (0.1 *M* NaCl)	3.91	43	5.12
$C_{12}H_{21}(OC_2H_4)_3SO_4^-Na^+$ (0.1 *M* NaCl)	3.52	52	5.34
$C_{12}H_{25}C_6H_4SO_3^-Na^+$	4.15	40	4.81
$C_9H_{19}C_6H_4(OC_2H_4)_4SO_4^-Na^+$	3.82	44	4.41
$C_9H_{19}C_6H_4(OC_2H_4)_4SO_4^-Na^+$	3.61	46	4.32
$C_9H_{19}C_6H_4(OC_2H_4)_9SO_4^-NH_4$	3.33	50	3.92
$C_{12}H_{25}C_6H_3(SO_3^-Na^+)OC_6H_4SO_3^-Na^+$	1.84	90	3.71
$C_4H_9CH(C_2H_5)OOCCH_2CH(SO_3^-Na^+)COOCH_2CH$ $(C_2H_5)C_4H_9$	1.52	105	4.15
$C_8H_{17}(OC_2H_4)_9O]_{1,2}P(O)(OH)_{1,2}$ (pH 3)	3.35	50	3.11
$C_{12}H_{25}(OC_2H_4)_4O]_{1,2}P(O)(OH)_{1,2}$ (pH 3)	3.02	55	5.85
$[C_{18}H_{35}(OC_2H_4)_4O]_{1,2}P(O)(OH)_{1,2}$ (pH 3)	2.52	66	5.80
$[C_9H_{19}C_6H_4(OC_2H_4)_4O]_{1,2}P(O)(OH)_{1,2}$ (pH 3)	2.48	67	5.42
$[C_9H_{19}C_6H_4(OC_2H_4)_4O]_{1,2}P(O)(O^-Na^+)_{1,2}$ (pH 7)	1.91	87	4.32
$C_{11}H_{23}C(O)N(CH_3)CH_2CH_2SO_3^-Na^+$	4.01	42	4.12
$C_{15}H_{31}C(O)N(CH_3)CH_2CH_2SO_3^-Na^+$	3.54	48	4.15
$C_{15}H_{31}C(O)N(C_6H_{11})CH_2CH_2SO_3^-Na^+$	3.15	53	4.35
$C_{17}H_{33}C(O)N(CH_3)CH_2CH_2SO_3^-Na^+$	3.02	55	4.27
(Tall oil acyl)$C(O)N(CH_3)CH_2CH_2SO_3^-Na^+$	3.25	51	3.30
$(C_2F_5)_{3-8}CH_2CH_2SCH_2CH_2CO_2^-Li^+$	3.61	46	4.62
Zwitterionics			
(br)*$C_8H_{17}N^+H(CH_2CH_2COO^-)CH_2CH_2C(O)OH$ (pH 6)	2.55	65	1.64
$C_9H_{19}C(O)NH(CH_2)_2N^+H(CH_2CH_2OH)CH_2COO^-$ (pH 6)	2.68	62	2.42
$C_{11}H_{23}C(O)NH(CH_2)_2N^+H(CH_2CH_2OH)CH_2COO^-$ (pH 6)	3.60	46	4.72
$C_{11}H_{23}C(O)NH(CH_2)_2N^+(CH_2CH_2OH)CH_2COOH$ (pH 6) | CH_2COO^-	3.32	50	4.72
$C_{11}H_{23}C(O)NH(CH_2)_3N^+(CH_3)_2CH_2COO^-$ (pH 6)	3.46	48	4.81
$C_{17}H_{33}C(O)NH(CH_2)_3N^+(CH_3)_2CH_2COO^-$ (pH 6)	3.11	53	4.91
$C_{12}H_{25}N^+H(CH_2CH_2COO^-)CH_2CH_2COOH$ (pH 6)	3.01	55	5.09
$C_{11}H_{23}CONH(CH_2)_3N^+(CH_3)_2CH_2CH(OH)CH_2SO_3^-$ (pH 6)	2.86	58	5.51
$(C_2F_5)_{3-8}CH_2CH(OCOCH_3)CH_2N^+(CH_3)_2CH_2CO_2^-$ (pH 6)	—	—	4.60

(continued)

TABLE 2.3
(Continued)

Surfactants	Γ_m(mol/ cm^2 × 10^{10})	A_m, $Å^2$	pC_{20}
Cationics			
$C_{12}H_{25}N^+(CH_3)_3Cl^-$	4.1	40	2.8
$C_{12}H_{25}N^+(CH_3)_3Br-$	3.9	43	3.0
$C_{12}H_{25}N^+(CH_2C_6H_5)(CH_3)_2Cl^-$	3.31	50	3.2
$C_{16}H_{33}N^+(CH_3)_3Cl^-$	2.71	61	—
$C_{16}H_{33}N^+(CH_3)_3Br-$	2.56	65	—
Zwitterionic/anionic salts (1:1 in 0.1 M NaCl)			
$C_{11}H_{23}C(O)NH(CH_2)_2N^+H(C_2H_4OH)CH_2COO^-$, $C_{12}H_{25}OC_2H_4SO_4^-$	3.95	42	5.37
$C_{11}H_{23}C(O)NH(CH_2)_2N^+H(C_2H_4OH)CH_2COO^-$, $C_{12}H_{25}(OC_2H_4)_2SO_4^-$	4.10	40	5.31
$C_{11}H_{23}C(O)NH(CH_2)_2N^+H(C_2H_4OH)CH_2COO^-$, $C_{12}H_{25}(OC_2H_4)_3SO_4^-$	4.10	40	5.35
$C_{11}H_{23}C(O)NH(CH_2)_3N^+(CH_3)_2CH_2COO-$, $C_{12}H_{25}OC_2H_4SO_4^-$	2.38	70	5.44
$C_{11}H_{23}C(O)NH(CH_2)_3N^+(CH_3)_2COO-$, $C_{12}H_{25}(OC_2H_4)_2SO_4^-$	2.55	65	5.65
$C_{11}H_{23}C(O)NH(CH_2)_3N^+(CH_3)_2CH_2COO-$, $C_{12}H_{25}(OC_2H_4)_3SO_4^-$	2.68	62	5.82

*Branched chain.

the greater the value of Γ_{max}, the larger will be the change in properties, other factors being constant.

The **rate** of adsorption of surfactants at a surface is generally determined from a dynamic surface tension plot—a plot of surface tension as a function of time. Commercial instruments are now available for measuring the surface tension of surfactant solutions as a function of time, in the range of 0.2–100 sec, by the maximum bubble pressure method. The rate at which surface tension is lowered is of great importance in high-speed industrial processes, such as dyeing and coating of surfaces.

Adsorption at Liquid/Liquid and Solid/Liquid Interfaces

Adsorption at the liquid/liquid interface can often be determined by use of a plot of interfacial tension (γ_I) vs. log C, in the same manner as for adsorption at a surface, by use of the Gibbs adsorption (Eq. 2.2). The slope of the γ_I–log C plot can be used to calculate the values of Γ at the

TABLE 2.4
Dynamic Surface Tension (mN/m) of Aqueous Solutions of Some Commercial Surfactants at Concentration 0.1% by Weight

Surfactants	Dynamic surface tension	
	1.0 sec	0.1 sec
Nonionics		
$C_{9-11}H_{21}(OC_2H_4)_6OH$	35.1	47.2
$C_{9-11}H_{21}(OC_2H_4)_8OH$	40.0	52.0
(br)*$C_{10}H_{21}(OC_2H_4)_4OH$	32.0	43.0
(br)*$C_{10}H_{21}(OC_2H_4)_6OH$	38.0	45.0
(br)*$C_{10}H_{21}(OC_2H_4)_{14}Cl$	30.5	38.5
$C_{11}H_{23}(OC_2H_4)_{10}OH$	52.0	65.0
$C_{11}H_{23}(OC_2H_4)_{10}Cl$	32.0	48.5
$C_{11}H_{23}(OC_2H_4)_{14}Cl$	36.0	51.0
$C_{11}H_{23}(OC_2H_4)_7OH$	42.0	60.0
$C_{12}H_{25}(OC_2H_4)_9OH$	49.0	60.5
$C_{12}H_{25}(OC_2H_4)_{12}OH$	51.5	60.5
p-t-$C_8H_{17}C_6H_4(OC_2H_4)_7OH$	32.5	50.0
p-t-$C_8H_{17}C_6H_4(OC_2H_4)_9OH$	34.5	50.5
p-t-$C_8H_{17}C_6H_4(OC_2H_4)_{12}OH$	41.0	56.0
p-t-$C_8H_{17}C_6H_4(OC_2H_4)_{30}OH$	55.5	62.0
p-t-$C_9H_{19}C_6H_4(OC_2H_4)_6OH$	52.0[a]	71.0[a]
p-t-$C_9H_{19}C_6H_4(OC_2H_4)_9OH$	35.2	55.5
p-t-$C_9H_{19}C_6H_4(OC_2H_4)_{12}OH$	45.0	62.0
$(C_9H_{19})_2C_6H_3(OC_2H_4)_9OH$	58.0	71.0
(tert) $C_9H_{19}C(CH_3)_2S(C_2H_4O)_6H$	40.0[a]	64.0[a]
(tert) $C_9H_{19}C(CH_3)_2S(C_2H_4O)_8H$	36.5	55.5
(tert) $C_9H_{19}C(CH_3)_2S(C_2H_4O)_{10}H$	38.5	56.0
$C_{13}H_{27}(OC_2H_4)_6OH$	41.0[a]	59.0[a]
$C_{13}H_{27}(OC_2H_4)_9OH$	34.0	51.0
$C_{8-10}H_{17-21}O(C_2H_4O)_5(C_3H_7O)_5H$	37.0	48.0
$C_{8-10}H_{17-21}O(C_2H_4O)_6(C_3H_7O)_5H$	34.5	44.0
$C_{8-10}H_{17-21}O(C_2H_4O)_6(C_3H_7O)_5H$[b]	31.5	38.0
(br)*$C_{10}H_{21}O(C_2H_4O)_5(C_3H_7O)_5H$	34.0	42.0
(br)*$C_{10}H_{21}O(C_2H_4O)_6(C_3H_7O)_5H$	36.0	45.0
$C_9H_{19}C_6H_4(OC_2H_4)_9(OC_3H_7)_{15}OH$	36.5	47.0
Nopol**$(C_3H_7O)_3(C_2H_4O)_5H$	35.0	45.0
Nopol $(C_3H_7O)_3(C_2H_4O)_6H$	38.0	50.0
Nopol $(C_3H_7O)_3(C_2H_4O)_7H$	38.0	49.0
Nopol $(C_3H_7O)_3(C_2H_4O)_9H$	45.0	54.0
$(CH_3)_3(SiO_2)_2Si(CH_3)(CH_2)_3(OC_2H_4)_{7.5}OCH_3$	28.0	57.0

(continued)

TABLE 2.4
(Continued)

Surfactants	Dynamic surface tension 1.0 sec	Dynamic surface tension 0.1 sec
Anionics		
$C_8H_{17}SO_4^-Na^+$	67.8	70.5
(br)*$C_8H_{17}SO_4^-Na^+$	72.8	79.2
$C_{12}H_{25}SO_4^-Na^+$	47.0	62.0
$C_{12}H_{25}SO_4^-NH^+_4$	38.0	59.0
$C_{12}H_{25}OC_2H_4SO_4^-Na^+$	43.0	59.0
$C_{12}H_{25}(OC_2H_4)_2SO_4^-Na^+$	50.0	61.0
$C_{12}H_{25}(OC_2H_4)_3SO_4^-Na^+$	53.0	62.0
$C_{13}H_{27}SO_4^-Na^+$	47.0	62.0
$C_{18}H_{35}SO_4^-Na^+$	44.0	69.0
$C_{12}H_{25}C_6H_4SO_3^-Na^+$	49.0	65.0
(br)*$C_{12}H_{25}C_6H_4SO_3^-Na^+$	42.0	58.0
$C_4H_9CH(C_2H_5)OOCCH_2CH(SO_{3-Na+})COOCH_2CH(C_2H_5)C_4H_9$	37.0	47.0
$C_9H_{19}OOCCH_2CH(SO_3^-Na^+)COOC_9H_{19}$	36.0	56.0
Sodium di-isopropyl naphthalene sulfonate	64.0	71.0
Sodium dibutyl naphthalene sulfonate	42.0	62.0
Sodium naphthalene formaldehyde sulfonate	73.0	75.0
$C_{12}H_{25}C_6H_3(SO_3^-Na^+)O\ C_6H_4SO_3^-Na^+$	63.0	71.0
$[C_{8-10}H_{17-21}(OC_2H_4)O]_{1,2}P(O)(O^-Na^+)_{1,2}$ (pH 7)	48.0	55.0
$[C_{13}H_{27}(OC_2H_4)_6O]_{1,2}P(O)(O^-Na^+)_{1,2}$ (pH 7)	52.0	65.0
$[C_9H_{19}C_6H_4(OC_2H_4)_4O]_{1,2}P(O)(O^-Na^+)_{1,2}$ (pH 7)	46.0	67.0
$[C_9H_{19}C_6H_4(OC_2H_4)_9O]_{1,2}P(O)(O^-Na^+)_{1,2}$ (pH 7)	53.0	67.0

liquid/liquid interface and values for the efficiency and effectiveness of adsorption at the liquid/liquid interface can be calculated in analogous fashion to that used at the surface.

Since interfacial tensions at the solid/liquid interface are generally difficult to measure, the use of Equation 2.2 to measure, indirectly, the concentration of surfactant at the solid/liquid interface is generally not feasible. Instead, direct determination of the amount adsorbed is made from the change in the concentration of the surfactant in the solution phase, before and after shaking it with the (finely divided) solid to equilibrium at constant temperature. From the change in the surfactant concentration (ΔC), the mass of the solid (m), and the volume of the surfactant solution (V), the amount of surfactant (n) adsorbed per unit weight of solid is (l[e])

TABLE 2.4
(Continued)

Surfactants	Dynamic surface tension 1.0 sec	0.1 sec
Cationics		
$C_{16}H_{37}N^+(CH_3)_3Br^-$	46.0	58.0
$C_{16}H_{37}N^+(CH_3)_3Cl^-$	40.0	58.0
Anionics		
(br)*$C_8H_{17}N^+H(CH_2CH_2COOH)CH_2CH_2COO^-$ (pH 6)	66.0	69.0
$C_9H_{19}C(O)NHCH_2CH_2NH^+(CH_2CH_2OH)CH_2COO^-$ (pH 6)	63.0	71.0
$C_{12}H_{25}N^+H(CH_2CH_2COOH)CH_2CH_2COO^-$ (pH 6)	48.0	67.0
$C_{11}H_{23}C(O)NHCH_2CH_2NH^+(CH_2CH_2OH)CH_2COO^-$ (pH 6)	42.0	50.5
$C_{11}H_{23}CONHCH_2CH_2N^+(CH_2CH_2OH)(CHvCOO^-)$ CH_2COOH (pH 6)	57.0	70.0
$C_{11}H_{23}C(O)NHCH_2CH_2CH_2N^+(CH_3)_2CH_2COO^-$ (pH 6)	47.0	61.0
$C_{17}H_{35}(O)NHCH_2CH_2CH_2N^+(CH_3)_2CH_2COO^-$ (pH 6)	50.5	71.0
$C_{11}H_{23}CONH(CH_2)_3N^+(CH_3)_2CH_2CH(OH)CH_2SO_3^-$(pH 6)	47.0	64.0
$C_{12}H_{25}$(lauryl)$N^+(CH_3)_2O^-$	42.0	53.0
$C_{12}H_{25}$(coco)$N^+(CH_3)_2O^-$	53.0	73.0

[a]Cloudy solution.
[b]Heteric.
*Branched chain.
**Nopol

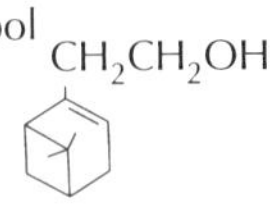

$$n = \Delta CV/m \quad [2.4]$$

where C is measured in mol/liter, V is in liters, and m is in grams. If the surface area, a^s, per gram of solid is known, n can be converted to mol/surface area of solid, Γ, by dividing it by a^s. The surface concentration of surfactant, Γ in mol/cm^2, is then:

$$\Gamma = \frac{\Delta C \times V}{m \times a^s} \quad [2.5]$$

where C is in mol/liter, V is in liters, m is in grams, and a^s is in cm^2/g. The area occupied by the molecule at the surface, in Å^2, is then $A = 10^{16}N/\Gamma$.

Adsorption data obtained at constant temperature in this fashion are usually plotted as a function of the equilibrium concentration of surfactant in the solution phase, yielding an **adsorption isotherm**. Adsorption isotherms of surfactants on solids, especially hydrophobic ones, are often of the Langmuir type (Fig. 2.5). From such an isotherm, the maximum concentration of surfactant at the interface (Γ_{max}, i.e., the *effectiveness* of adsorption of the surfactant) can be read directly from the plot, and the minimum area (A_{min}) occupied by the surfactant there calculated by Equation 2.3. The surfactant concentration in the solution phase required to produce that maximum concentration of surfactant at the interface, i.e., the *efficiency* of adsorption of the surfactant, can also be read directly from the adsorption isotherm.

The amount of reduction of the interfacial tension, π, at the solid/liquid interface caused by adsorption there of the surfactant up to the critical micelle concentration (cmc) can be determined from the plot of Γ vs. log (or ln) C (Fig. 2.5) by integrating the area under the curve since, from the Gibbs adsorption equation (2.2), $-d\gamma = 2.30nRT\Gamma d \log C$, for values up to the cmc.

In some cases, the equilibrium **contact angle** of a drop of surfactant solution on a smooth, planar, nonporous solid substrate (Fig. 2.6) can be used to indirectly determine the concentration of a surfactant adsorbed at that solid/liquid interface. The relationship among the contact angle, θ, of a surfactant solution at constant temperature on a smooth, planar substrate, the surface tension (or surface free energy), γ_{SA}, of that solid substrate, the interfacial tension, γ_{LS}, at the liquid/solid interface, and the sur-

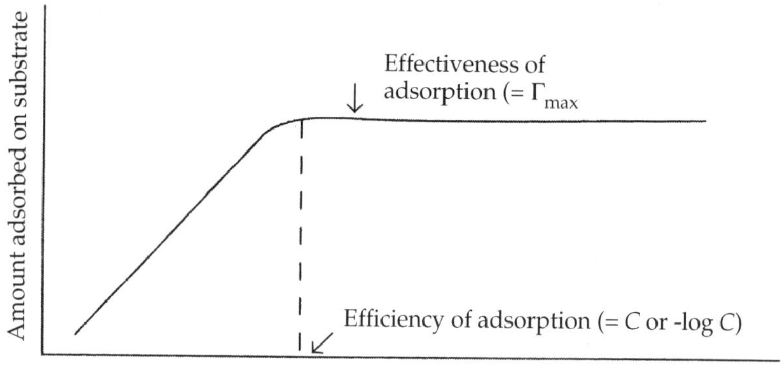

Figure 2.5. Langmuir-type adsorption isotherm, illustrating efficiency and effectiveness of adsorption on a solid substrate.

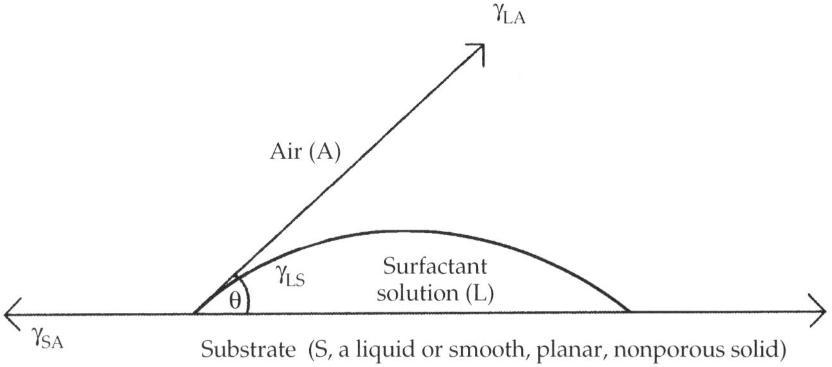

Figure 2.6. Contact angle of surfactant solution on smooth, planar, nonporous solid, illustrating relationship of various interfacial tensions.

face tension, γ_{LA}, of the surfactant solution is given by the equation (1[f]):

$$\gamma_{LA}\cos\theta = \gamma_{SA} - \gamma_{LS} \qquad [2.6]$$

Some contact angles of 0.1% aqueous solutions on Parafilm M at 22°C are listed in Table 2.5. The lowest contact angles are given by nonionic surfactants. Branched hydrophobes appear to give lower contact angles than isomeric linear ones.

If the solid has a nonpolar, hydrophobic (i.e., low energy) surface, then it can be assumed that adsorption of the surfactant at the solid/air surface will not occur to any significant extent (since that would *increase* its surface tension), and that consequently γ_{SA} will remain constant.* Under that condition, the value of Γ_{LS}, the molar concentration of surfactant at the solid/liquid interface, is given by the relationship:

$$\gamma_{LS} = [\partial(\gamma_{LA}\cos\theta)/\partial \log C]_T/2.303nRT \qquad [2.7]$$

with n, R, and T retaining their meanings from Equation 2.2, shown previously. Here, data on the change in the contact angle and the surface tension of the surfactant solution with change in the surfactant concentration in the solution phase permit calculation of the surfactant concentration at the solid/liquid interface.

*This is not true for fluorocarbon or polysiloxane-based surfactants on hydrocarbon surfaces, since their hydrophobic groups can reduce the surface free energy of hydrocarbon surfaces.

TABLE 2.5
Contact Angles of Surfactant Solutions at 0.1% on Parafilm M[a] (22°C)

Surfactants	Contact angle (°)
Nonionics	
$C_{9-11}H_{19-23}(OC_2H_4)_6OH$	26
$C_{9-11}H_{19-23}(OC_2H_4)_8OH$	29
$C_{9-11}H_{19-23}(OC_2H_4)_9OH$	30
(br)*$C_{10}H_{21}(OC_2H_4)_4OH$	7
(br)*$C_{10}H_{21}(OC_2H_4)_6OH$	5
(br)*$C_{11}H_{21}(OC_2H_4)_{10}Cl$	12
(br)*$CC_{11}H_{21}(OC_2H_4)_{14}Cl$	20
$C_{12}H_{25}(OC_2H_4)_7OH$	18
$C_{12}H_{25}(OC_2H_4)_9OH$	27
$C_{12-15}H_{25-31}(OC_2H_4)_7OH$	19
$C_{12-13}H_{25-27}(OC_2H_4)_7OH$	20
$C_{12-15}H_{25-31}(OC_2H_4)_9OH$	20
(br)*$C_{13}H_{27}(OC_2H_4)_7OH$	17
(br)*$C_{13}H_{27}(OC_2H_4)_{14}OH$	20
$C_{18}H_{35}(OC_2H_4)_{20}OH$	48
p-t-$C_8H_{17}C_6H_4(OC_2H_4)_5OH$	14
p-t-$C_8H_{17}C_6H_4(OC_2H_4)_7OH$	16
p-t-$C_8H_{17}C_6H_4(OC_2H_4)_9OH$	29
p-t-$C_8H_{17}C_6H_4(OC_2H_4)_{12}OH$	63
p-t-$C_8H_{17}C_6H_4(OC_2H_4)_{30}OH$	75
p-t-$C_9H_{19}C_6H_4(OC_2H_4)_6OH$	26
p-t-$C_9H_{19}C_6H_4(OC_2H_4)_9OH$	29
p-t-$C_9H_{19}C_6H_4(OC_2H_4)_{11}OH$	24
p-t-$C_9H_{19}C_6H_4(OC_2H_4)_{15}OH$	42
p-t-$C_9H_{19}C_6H_4(OC_2H_4)_{20}OH$	50
p-t-$C_9H_{19}C_6H_4(OC_2H_4)_{30}OH$	55
$C_{8-10}H_{17-21}O(C_2H_4O)_4(C_3H_7O)_5H$	16
(br)*$C_{10}H_{21}O(C_2H_4O)_4(C_3H_7O)_5H$	18
$(C_2H_4O)_3(C_3H_7O)_3(C_2H_4O)_3$	70
$(C_2H_4O)_8(C_3H_7O)_{30}(C_2H_4O)_8$	68
Castor oil$(OC_2H_4)_{30}OH$	46
Castor oil$(OC_2H_4)_{40}OH$	71
(br)*$C_{12}H_{25}S(C_2H_4O)_6H$	13
(br)*$C_{12}H_{25}S(C_2H_4O)_7H$	12
Anionics	
$C_8H_{17}SO_4^-Na^+$	60
(br)*$C_8H_7SO_4^-Na^+$	63
$C_{12}H_{25}SO_4^-Na^+$	56
$C_{12}H_{25}SO_4^-NH^+{}_4$	45

(*continued*)

TABLE 2.5
(Continued)

Surfactants	Contact angle (°)
Anionics (continued)	
$C_{12}H_{21}OC_2H_4SO_4^-Na^+$	26
$C_{12}H_{25}(OC_2H_4)_2SO_4^-Na^+$	45
$C_{12}H_{21}(OC_2H_4)_3SO_4^-Na^+$	53
$C_4H_9CH(C_2H_5)OOCCH_2CH(SO_3^-Na^+)COOCH_2CH(C_2H_5)C_4H_9$	47
$C_9H_{19}OOCCH_2CH(SO_3^-Na^+)COOC_9H_{19}$	29
$[C_8H_{17}(OC_2H_4)_6O]_{1,2}P(O)(O^-Na^+)_{1,2}$ (pH 7)	43
$[C_{10}H_{21}(OC_2H_4)_4O]_{1,2}P(O)(O^-Na^+)_{1,2}$ (pH 7)	40
$[C_9H_{19}C_6H_4(OC_2H_4)_4O]_{1,2}P(O)(O^-Na^+)_{1,2}$ (pH 7)	50
$[(br)^*C_9H_{19}C_6H_4(OC_2H_4)_6O]_{1,2}P(O)(O^-Na^+)_{1,2}$ (pH 7)	58
$[(br)^*C_{13}H_{27}C_6H_4(OC_2H_4)_6O]_{1,2}P(O)(O^-Na^+)_{1,2}$ (pH 7)	54
Cationics	
$C_{12}H_{25}N^+(CH_3)_2CH_2C_6H_5Cl^-$	63
$C_{16}H_{25}N^+(CH_3)_3Cl^-$	62
$C_{16}H_{25}N^+(CH_3)_3Br^-$	59
$C_{16}H_{33}N^+(C_3H_7)_3Br^-$	61
$C_{18}H_{35}N^+(CH_3)_3Br^-$	58
Zwitteronics[b]	
$(br)^*C_8H_{17}N(CH_2CH_2COO^-Na^+)_2$ (pH 8.0)	62
$C_9H_{19}C(O)NHCH_2CH_2N^+H(CH_2CH_2OH)CH_2COO^-$ (pH 6.0)	50
$C_{12}H_{25}N^+H(CH_2CH_2COOH)CH_2CH_2COO^-$ (pH 6.0)	35
$C_{11}H_{23}CONHCH_2CH_2N^+(CH_2CH_2OH)(CH_2COOH)CH_2CH_2COO^-$ (pH 6)	39
$C_{17}H_{35}C(O)NHCH_2CH_2CH_2N^+(CH_3)_2CH_2COO^-$ (pH 6)	43
$C_{11}H_{23}CONH(CH_2)_3N^+(CH_3)_2CH_2CH(OH)CH_2SO_3^-$ (pH 6.0)	48

[a]Available from Fisher Scientific, Pittsburgh, PA.
[b]May be anionic at pH >7.
*Branched chain.

Again, the area occupied at the interface by the surfactant molecule can be calculated by use of Equation 2.3.

Changes in Performance Phenomena Resulting from Surfactant Adsorption

Adsorption at the Surface of Aqueous Solutions of Surfactants

As mentioned above (in Changes in the Properties of the Surface of a Solution, and Changes in the Properties at Solid/Liquid and Liquid/Liquid Interfaces), surfactant adsorption at the interface between two phases results in a change in the interfacial tension between them. If the adsorbed surfactant is oriented with its hydrophobic group toward the more nonpolar phase and with its hydrophilic group toward the more polar phase, then the interfacial tension between the two phases will be reduced; if the orientation is with the hydrophilic group toward the more nonpolar phase and the hydrophobic group toward the more polar phase, then the tension between them will be increased. If the interfacial tension is reduced, it will be easier to increase the interfacial area between the two phases; if it is increased, it will be more difficult to increase that interface (or the interface will have a greater tendency to contract). Some performance phenomena directly affected by the interfacial tension between two phases are wetting and dewetting; foaming and defoaming; emulsification and demulsification; dispersion and flocculation of solid; and adhesion.

Wetting and Dewetting. Wetting, in its most general sense, is the displacement of one fluid from its interface with a substrate by another fluid. In the case of surfactant utilization, the most usual cases are when a surfactant solution (L) is used to displace a liquid, generally oily (O), from its interface with a substrate (Fig. 2.7a), or when it is used to displace air (A) from the surface of a substrate (Fig. 2.7b). γ_{SO} is the interfacial tension between the substrates and the liquid, O; γ_{LS} is the interfacial tension between the surfactant solution, L, and the substrate; γ_{LO} is

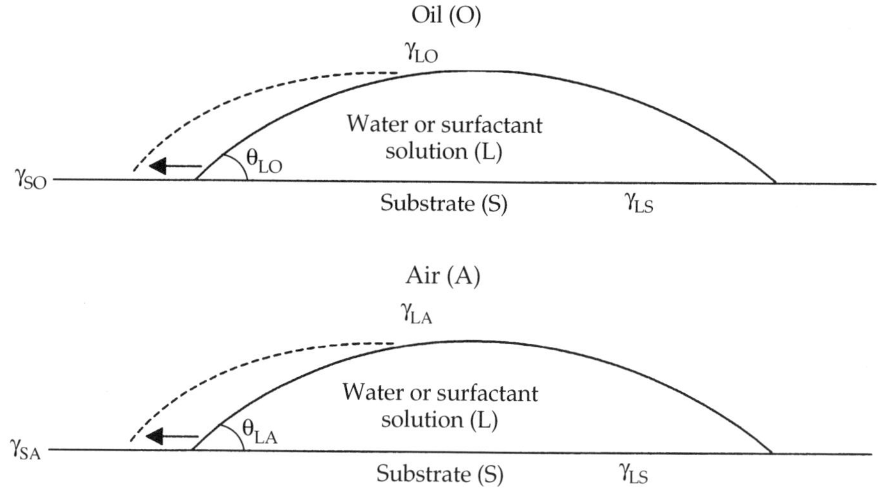

Figure 2.7. (a) Wetting of a substrate by a liquid, displacing an oily substance; (b) wetting of a substrate by a liquid, displacing air.

the interfacial tension between L and O; γ_{SA} is the surface tension of the substrate; and γ_{LA} is the surface tension of the surfactant solution.

A useful measure of the tendency of L to spontaneously spread over S is the spreading coefficient, $S_{L/S}$, given by the relationship among the three interfacial tensions involved:

For Fig. 2.7a,

$$S_{L/S} = \gamma_{SO} - \gamma_{LS} - \gamma_{LO} \quad [2.8]$$

For Fig. 2.7b,

$$S_{L/S} = \gamma_{SA} - \gamma_{LS} - \gamma_{LA} \quad [2.9]$$

The more positive the value of the spreading coefficient, the greater the tendency to wet the substrate. Therefore, decrease in the surface tension of the wetting liquid and decrease in the interfacial tension of the wetting liquid against the substrate to be wet both increase its wetting tendency.

When the surfactant adsorbs at the interface, either with the substrate or with the second fluid (O or A), in such orientation as to decrease the interfacial tensions there, wetting will be facilitated. If, on the other hand, it adsorbs with such orientation as to increase those tensions, then it will be more difficult to wet them (i.e., dewetting will occur).

If the substrate is nonpolar and hydrophobic, then adsorption of the surfactant onto it from an aqueous solution will be with its hydrophobic group oriented toward the hydrophobic substrate and its hydrophilic group toward the water. This will decrease γ_{LS}, making $S_{L/S}$ more positive, and therefore increasing the tendency of the surfactant solution to spontaneously wet the substrate.

The surfactant will also adsorb at the oily liquid, O, or air (A) interface, with its hydrophobic group toward that second fluid and its hydrophilic group oriented toward the water. That will decrease γ_{LO} or γ_{LA}, making $S_{L/S}$ even more positive and increasing even more the tendency of the surfactant solution to wet the substrate.

When the substrate is nonporous and sufficient time (a few minutes or more) exists for equilibrium conditions to be established, then the surfactant solution (L) will have an equilibrium contact angle, θ (measured in the surfactant solution), against the other two phases.

The relationship between the contact angle, θ, and the interfacial tensions in the case of Figure 2.7a, is:

$$\gamma_{LO} \cos\theta_{LO} = \gamma_{SO} - \gamma_{LS} \quad [2.10]$$

from which

$$S_{L/S} = \gamma_{LO} (\cos \theta_{LO} - 1) \quad [2.11]$$

In the case of Fig. 2.7b,

$$\gamma_{LA} \cos \theta_{LA} = \gamma_{SA} - \gamma_{LS} \quad [2.12]$$

from which

$$S_{L/S} = \gamma_{LA}(\cos \theta_{LA} - 1) \quad [2.13]$$

The smaller the contact angle, θ_{LA}, that the surfactant solution makes with the substrate and the smaller the interfacial tension, γ_{LO}, between it and the oily liquid, the more positive will be the spreading coefficient and consequently the tendency of the surfactant solution to displace O from S. Again, adsorption of the surfactant at the interfaces with the oily liquid and the substrate with such orientation as to reduce the tensions at those interfaces will increase the tendency of the surfactant solution to displace the oily liquid from the substrate.

The Draves cotton skein wetting test is commonly used to measure the rate at which aqueous solutions of surfactants can wet a porous, partially hydrophobic substrate (3). Table 2.6 lists the weight concentrations of various commercial surfactants required to obtain a 25-sec wetting time in this test. The lowest concentrations are with surfactants that have branched, rather than isomeric, linear hydrophobes. A decrease in the polarity, or ionic nature, of the hydrophilic group, provided that the surfactant remains completely in solution in the aqueous medium, also decreases the concentration required for 25-sec wetting. A correlation has been found between Draves cotton skein wetting time and dynamic surface tension at 1 sec surface age (4).

From Equations 2.12 and 2.13, it can be seen that the tendency of the surfactant solution to displace the second fluid, O or A, and wet the substrate, S, can be determined by measuring the interfacial tension of the surfactant solution against the second fluid and the contact angle, θ, that it makes with the substrate in the presence of the second fluid. In the case shown in Figure 2.7b, wetting a substrate by displacing air from it, this involves merely measuring the surface tension of the surfactant solution and the contact angle that it makes with the substrate.

In the case shown in Figure 2.7a, displacing a liquid, O, from the interface with the substrate, determining the value of the spreading coefficient is somewhat more difficult since it is usually difficult or inconvenient to

TABLE 2.6
Concentration (% by Weight) of Surfactants to Give 25-Sec Wetting Time (Draves Cotton Skein Wetting)

Surfactants	25-sec wetting time 22°C	70°C
Nonionics		
(br)*$C_{10}H_{21}(OC_2H_4)_4OH$	0.027	—
(br)*$C_{10}H_{21}(OC_2H_4)_6OH$	0.038	—
$C_{11}H_{23}(OC_2H_4)_{10}Cl$	0.053	0.051 (60°C)
$C_{11}H_{23}(OC_2H_4)_{14}Cl$	0.077	0.041 (60°C)
$C_{8-10}H_{17-21}O(C_2H_4O)_4(C_3H_7O)_5H$	0.056	0.064 (60°C)
$C_{8-10}H_{17-21}O(C_2H_4O)_6(C_3H_7O)_5H$	0.064	0.056 (60°C)
(br)*$C_{12-14}H_{25-29}(OC_2H_4)_7OH$ (20 sec)	046	—
(br)*$C_{12-14}H_{25-29}(OC_2H_4)_9OH$ (20 sec)	0.054	—
(br)*$C_{12-14}H_{25-29}(OC_2H_4)_{12}OH$ (20 sec)	0.068	—
(br)*$C_{12-14}H_{25-29}(OC_2H_4)_{15}OH$ (20 sec)	0.073	—
$C_{13}H_{27}(OC_2H_4)_9OH$	0.059	0.045
$C_{13}H_{27}(OC_2H_4)_{15}OH$	0.260	0.082
p-t-$C_8H_{17}C_6H_4(OC_2H_4)_5OH$	0.055	0.030
p-t-$C_8H_{17}C_6H_4(OC_2H_4)_7OH$	0.048	0.035
p-t-$C_8H_{17}C_6H_4(OC_2H_4)_9OH$	0.038	0.027
p-t-$C_8H_{17}C_6H_4(OC_2H_4)_{12}OH$	0.110	0.030
$C_9H_{19}C_6H_4(OC_2H_4)_8OH$	0.055	0.044
$C_9H_{19}C_6H_4(OC_2H_4)_9OH$	0.040	0.028
$C_9H_{19}C_6H_4(OC_2H_4)_{10}OH$	0.054	0.045
$C_9H_{19}C_6H_4(OC_2H_4)_{11}OH$	0.062	0.045
$C_9H_{19}C_6H_4(OC_2H_4)_{12}OH$	0.078	0.042
$C_9H_{19}C_6H_4(OC_2H_4)_{15}OH$	0.148	0.085
Anionics		
$C_{11}H_{23}C(O)OCH_2CH_2SO_3^-Na^+$	—	0.041
$C_{13}H_{27}C(O)N(CH_3)CH_2CH_2SO_3^-Na^+$	—	0.051
$C_{15}H_{31}C(O)N(CH_3)CH_2CH_2SO_3^-Na^+$	—	0.059
$C_{17}H_{35}C(O)N(CH_3)CH_2CH_2SO_3^-Na^+$	—	0.068
$C_9H_{19}C_6H_4(OC_2H_4)_4SO_3^-Na^+$	0.051	0.048
$C_9H_{19}C_6H_4(OC_2H_4)_4SO_4^-NH_4^+$	0.059	0.045
$C_9H_{19}C_6H_4(OC_2H_4)_9SO_4^-NH_4^+$	0.109	0.057
$[C_8H_{17}(OC_2H_4)_6O]_{1,2}P(O)(OH)_{1,2}$ (pH 2)	0.083	0.054 (65°C)
$[C_8H_{17}(OC_2H_4)_6O]_{1,2}P(O)(O^-Na^+)_{1,2}$ (pH 7)	0.088	0.056 (65°C)
$[C_8H_{17}(OC_2H_4)_6O]_{1,2}P(O)(O^-Na^+)_{1,2}$ (pH 10)	0.084	0.054 (65°C)
$[C_{10}H_{21}(OC_2H_4)_4O]_{1,2}P(O)(OH)_{1,2}$ (pH 2)	0.252	0.139
$[C_{10}H_{21}(OC_2H_4)_4O]_{1,2}P(O)(O^-Na^+)_{1,2}$ (pH 10)	0.271	—
$[C_{13}H_{27}(OC_2H_4)_4O]_{1,2}P(O)(OH)_{1,2}$ (pH 3)	0.092	—
$[C_{13}H_{27}(OC_2H_4)_4O]_{1,2}P(O)(O^-Na^+)_{1,2}$ (pH 7)	0.101	—

(continued)

TABLE 2.6
(Continued)

Surfactants	25 sec-wetting time 22°C	70°C
Anionics (continued)		
$[C_9H_{19}C_6H_4(OC_2H_4)_4O]_{1,2}P(O)(OH)_{1,2}$ (pH 3)	0.250	—
$[C_9H_{19}C_6H_4(OC_2H_4)_4O]_{1,2}P(O)(O^-Na^+)_{1,2}$ (pH 10)	0.311	—
Cationics		
$C_{12}H_{25}N^+(CH_3)_3Cl^-$	>0.500	—
$C_{12}H_{25}Pyr^+Cl^-$	>0.500	—
$C_{14}H_{29}N^+(CH_3)_3Br^-$	>0.500	—
$C_{16}H_{25}N^+(CH_3)_3Br^-$	>0.500	—

*Branched chain.

measure the contact angle, θ_{DO}, of the surfactant solution on the substrate in the presence of the second fluid, O. In that event, the value of the spreading coefficient can be determined by measuring the contact angle of the second fluid by itself against air on the surface of the substrate (Fig. 2.8).

Since, from Figure 2.8,

$$\gamma_{OA} \cos \theta_{OA} = \gamma_{SA} - \gamma_{SO} \qquad [2.14]$$

combining Equations 2.8, 2.12, and 2.14 yields

$$S_{L/S} = \gamma_{LA} \cos \theta_{LA} - \gamma_{OA} \cos \theta_{OA} - \gamma_{LO} \qquad [2.15]$$

Equation 2.15 states that the spreading coefficient in this case can be determined by measuring the surface tensions of both the surfactant solution, L, and the liquid, O, by themselves, their respective contact angles on the substrate, S, and the interfacial tension between them.

If the substrate to be wet has strongly polar or ionic surface sites, however, then adsorption of the surfactant at the substrate/water interface

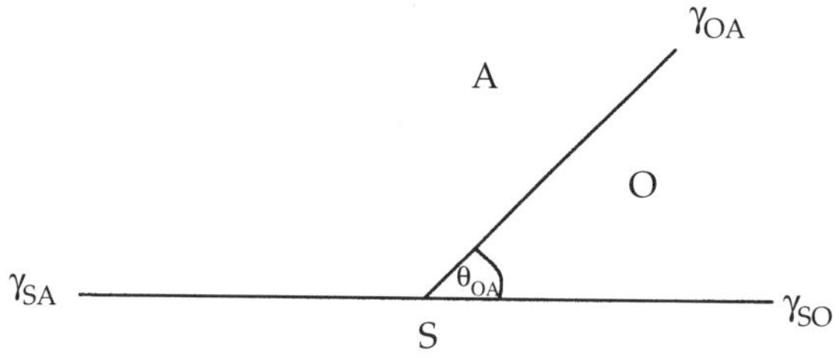

Figure 2.8. Contact angle of liquid (O) against air (A) on the surface of the substrate (S).

may occur with its hydrophilic head oriented toward the polar or ionic sites and its hydrophobic group oriented toward the water. This will increase the value of γ_{LS} and make wetting more difficult. Behavior of this sort is often encountered when using ionic surfactants to wet substrates that carry a substantial electrical charge when contacted with water or an aqueous solution. If the charge is negative, for example, then anionic surfactants will not adsorb significantly at the substrate/water interface because of the repulsion between the negative surfactant head and the negative substrate. As a result, any enhancement of wetting will be due only to reduction in the value of γ_{LO} or γ_{LA}. On the other hand, if a cationic surfactant is used, then it will be strongly adsorbed at the substrate/water interface because of the attraction between the positive head group of the surfactant and the negative substrate. The orientation of the adsorbed surfactant will be with its hydrophobic group toward the water, and the resulting increase in the value of $\gamma\Lambda_{LS}$ will now make the substrate much less wettable than in the absence of the cationic surfactant (see Dewetting later in this chapter). In addition, the original electrical charge of the substrate will have now been partially or completely neutralized by the adsorption onto it of the oppositely charged surfactant. Adsorption in this fashion of oppositely charged surfactants onto an originally charged substrate is the basis for some waterproofing treatments, and for hair "conditioning" and fabric "softening" treatments.

When time is insufficient for equilibrium conditions to be established in the wetting process, e.g., high-speed wetting of substrates, especially porous ones, then the tendency of the surfactant solution to wet the substrate depends on the rates of adsorption; that is, how rapidly adsorption occurs at the aqueous/substrate and aqueous/air interfaces to reduce the tensions there and consequently facilitate wetting. In many cases, equilibrium surface and interfacial tension values are not indicative of values at short times, and only dynamic values (e.g., surface or interfacial tension values at times of 1 sec or less) correlate with wetting performance.

In dewetting, the nature of the water/substrate interface (LS) is changed in order to facilitate the removal of water from a substrate. In the dewetting process (Fig. 2.9), new areas of liquid/air interface (LA) and substrate/air interface (SA) are formed and the liquid/substrate interface (LS) disappears.

The work required to remove the liquid, per unit area of interface, consequently is (from Eq. 2.1), $\gamma_{SA} + \gamma_{LA} - \gamma_{LS}$. By adsorbing at the LS interface in such manner as to increase γ_{LS}, and by adsorbing at the LA

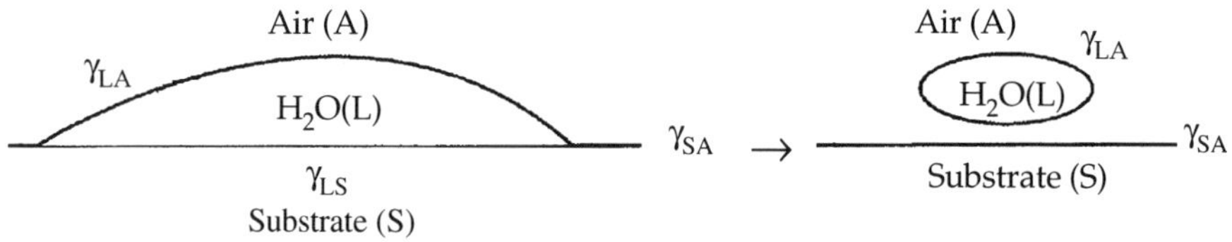

Figure 2.9. Dewetting: removal of water from the substrate.

interface in such manner as to decrease γ_{LA}, the surfactant decreases the work required to remove the water from the substrate. Therefore, surfactants that adsorb onto the substrate in such manner as to increase the hydrophobicity of the substrate (and consequently, increase (γ_{LS}) will facilitate the removal of water from it.

Foaming and the Reduction of Foaming. In foaming, a gas (usually air) is introduced beneath the surface of a liquid, which produces a dispersion of the gas in the liquid (Fig. 2.10a and b). The gas bubbles in the foam are each surrounded by a liquid film. This liquid film thins as gravity and the lower internal pressure in the Plateau region of higher curvature (where bubbles meet each other) cause the liquid to drain back into the bulk liquid phase. If there is no stabilizing mechanism, the liquid film will thin to the point at which it breaks and the gas bubbles

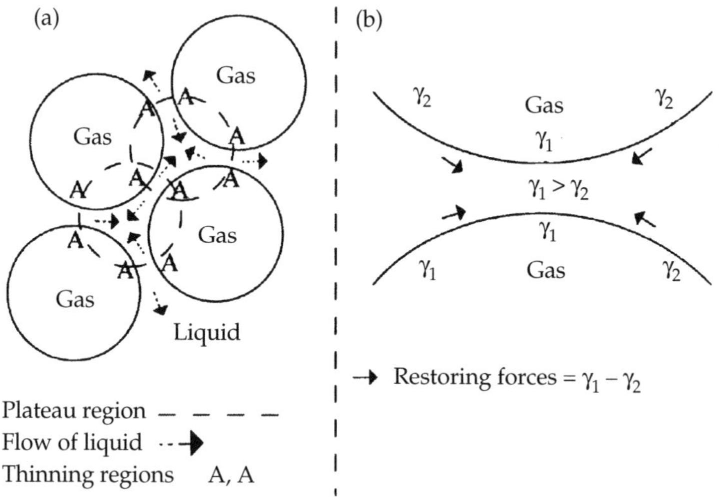

Figure 2.10. Gas bubbles in a foam: (a) showing plateau regions, thinning regions, flow of liquid phase into plateau regions; (b) magnification of thinning region, A, showing restoring flows due to $\gamma_1 > \gamma_2$.

coalesce and escape from the liquid. The presence of some solute that adsorbs at the gas/liquid interface can provide this destabilizing mechanism. The adsorbed solute decreases the surface tension of the liquid in the liquid film. When the liquid film starts to thin, the surface tension of the thinning portion momentarily increases for two reasons: (i) the concentration of adsorbed solute decreases as the area of the thinning portion stretches and increases, resulting in an increase in the surface tension of the thinning portion, and (ii) the surface tension at a new surface is always larger than the equilibrium value, because of the time needed for the solute to adsorb there. This difference in surface tension between the thinning portion of the liquid film and the thicker portions adjacent to it produces a restoring force—"film elasticity"—which resists further thinning and also draws liquid into it from the adjacent portions (Fig. 2.10b).

Foaming, therefore, depends on the presence in the foaming solution of some surface-active material that can produce differences in surface tension in the air/solution interface as that interface expands to enclose the gas bubbles.

Since a very large increase occurs in the area of the gas/liquid interface during aqueous foam formation, a reduction in the surface tension of the water (see Eq. 2.1) makes it easier for foaming to occur. The stability of the foam depends also on other features, however, such as the packing of surfactant molecules in the film (A_{min}) and the internal properties of the aqueous phase (Chapter 3). A closely packed layer of surfactant at the air/water interface generally improves the stability of the foam.

The Ross-Miles foaming test is an American Society for Testing Materials standard method for assessing foaming ability (5). Some foam heights of aqueous solutions of commercial surfactants, obtained by this method are listed in Table 2.7. Nonionic surfactants generally show poorer foam stability than anionic or zwitterionic surfactants. Cationics show poor foaming in this test, possibly because they adsorb onto the glass wall of the foaming apparatus *via* their positively charged hydrophilic head groups. Surfactants with linear hydrophobes show better foaming properties than their isomers with branched hydrophobes. Mixtures of zwitterionics capable of accepting a proton and anionics can show excellent foaming properties.

On the other hand, when little air or no foaming is desired in a surfactant-containing system, this condition can also be attained by use of surfactant adsorption at the surface of the aqueous system. One method is by reducing the film elasticity, described in this section, by use of a surfactant

TABLE 2.7
Foaming Heights of Aqueous Surfactant Solutions (Ross-Miles Method)

Surfactants	C (%)	Temp 0°C	Foam (mm) Initial	Final (5 min)
Nonionics				
$C_{8-10}H_{17-21}(C_2H_4O)_5C_3H_7O)_3H$	0.1	22	75	12
$C_{8-10}H_{17-21}(C_2H_4O)_5C_3H_7O)_3H$	0.1	50	4	0
$C_{8-10}H_{17-21}(C_2H_4O)_5C_3H_7O)_5H$	0.1	22	39	5
$C_{9-11}H_{19-23}(OC_2H_4)_6OH$	0.1	50	109	28
$C_{9-11}H_{19-23}(OC_2H_4)_6OH$	1.0	50	158	11
$C_{9-11}H_{19-23}(OC_2H_4)_8OH$	0.1	50	162	32
$C_{9-11}H_{19-23}(OC_2H_4)_8OH$	1.0	50	185	25
(br)*$C_{10}H_{21}(OC_2H_4)_4OH$	0.1	22	25	4
(br)*$C_{10}H_{21}(OC_2H_4)_4OH$	0.1	50	15	2
(br)*$C_{10}H_{21}(OC_2H_4)_6OH$	0.1	22	130	35
(br)*$C_{10}H_{21}(OC_2H_4)_6OH$	0.1	50	35	6
$C_{11}H_{23}(OC_2H_4)_5OH$	0.1	22	120	10
$C_{11}H_{23}(OC_2H_4)_7OH$	0.1	22	130	15
$C_{11}H_{23}(OC_2H_4)_{10}Cl$	0.1	22	13	5
(br)*$C_{11}H_{23}(OC_2H_4)_{10}Cl$	0.1	22	9	0
$C_{12}H_{25}(OC_2H_5)_7OH$	0.1	22	114	109
$C_{12}H_{25}(OC_2H_4)_{12}OH$	0.1	22	117	99
$C_{12-13}H_{25-27}(OC_2H_4)_7OH$	0.1	50	90	20
$C_{12-13}H_{25-27}(OC_2H_4)_7OH$	1.0	50	160	12
$C_{12-14}H_{25-29}(OC_2H_4)_7OH$	0.2	50	65	20
$C_{12-14}H_{25-29}(OC_2H_4)_9OH$	0.2	50	165	10
$C_{12-14}H_{25-29}(OC_2H_4)_{12}OH$	0.2	50	175	20
$C_{12-14}H_{25-29}(OC_2H_4)_{15}OH$	0.2	50	187	12
$C_{12-15}H_{25-31}(OC_2H_4)_7OH$	0.1	50	76	10
$C_{12-15}H_{25-31}(OC_2H_4)_7OH$	1.0	50	120	10
$C_{12-15}H_{25-31}(OC_2H_4)_9OH$	0.1	50	120	11
$C_{12-15}H_{25-31}(OC_2H_4)_9OH$	1.0	50	170	10
$C_{12-15}H_{25-31}(OC_2H_4)_{12}OH$	0.1	50	140	10
$C_{12-15}H_{25-31}(OC_2H_4)_{12}OH$	1.0	50	180	15
$C_{12-14}H_{25-31}(OC_2H_4)_{20}OH$	0.2	50	171	25
$C_{14-15}H_{29-31}(OC_2H_4)_7OH$	0.1	50	95	82
$C_{14-15}H_{29-31}(OC_2H_4)_7OH$	1.0	50	60	10
$C_{14-15}H_{29-31}(OC_2H_4)_{13}OH$	0.1	50	130	70
$C_{14-15}H_{29-31}(OC_2H_4)_{13}OH$	1.0	50	130	30
p-t-$C_8H_{17}C_6H_4(OC_2H_4)_5OH$	0.1	22	15	11
p-t-$C_8H_{17}C_6H_4(OC_2H_4)_7OH$	0.1	22	42	40
p-t-$C_8H_{17}C_6H_4(OC_2H_4)_7OH$	0.1	50	29	27
p-t-$C_8H_{17}C_6H_4(OC_2H_4)_9OH$	0.1	22	100	30

(continued)

TABLE 2.7
(Continued)

Surfactants	C (%)	Temp 0°C	Foam (mm) Initial	Foam (mm) Final (5 min)
Nonionics (continued)				
p-t-$C_8H_{17}C_6H_4(OC_2H_4)_9OH$	0.1	50	130	4
p-t-$C_8H_{17}C_6H_4(OC_2H_4)_{12}OH$	0.1	25	125	50
p-t-$C_9H_{19}C_6H_4(OC_2H_4)_5OH$	0.1	25	8	5
p-t-$C_9H_{19}C_6H_4(OC_2H_4)_6OH$	0.1	25	15	10
p-t-$C_9H_{19}C_6H_4(OC_2H_4)_7OH$	0.1	25	55	45
p-t-$C_9H_{19}C_6H_4(OC_2H_4)_8OH$	0.1	22	69	55
p-t-$C_9H_{19}C_6H_4(OC_2H_4)_9OH$	0.1	22	73	66
p-t-$C_9H_{19}C_6H_4(OC_2H_4)_9OH$	0.1	50	75	67
p-t-$C_9H_{19}C_6H_4(OC_2H_4)_{10}OH$	0.1	25	82	75
p-t-$C_9H_{19}C_6H_4(OC_2H_4)_{11}OH$	0.1	25	110	80
p-t-$C_9H_{19}C_6H_4(OC_2H_4)_{15}OH$	0.1	25	130	110
$(C_9H_{19})_2C_6H_3(OC_2H_4)_4OH$	0.1	25	15	10
$(C_9H_{19})_2C_6H_3(OC_2H_4)_{15}OH$	0.1	25	40	35
$(C_9H_{19})_2C_6H_3(OC_2H_4)_{25}OH$	0.1	25	50	30
Castor oil $(OC_2H_4)_{30}OH$	0.1	25	69	56
Castor oil $(OC_2H_4)_{40}OH$	0.1	25	35	32
$C_{18}H_{35}N[(C_2H_4O)_{2.5}]_2H$	0.1	22	43	42
$C_{18}H_{35}N[C_2H_4O)_{4.5}]_2H$	0.1	22	98	80
$(C_2F_5)_{3-8}CH_2CH_2O(C_2H_4O)_6H$	0.1	41	145	142
$(C_2F_5)_{3-8}CH_2CH_2O(C_2H_4O)_6H$	0.1	68	130	60
$C_9H_{19}C(CH_3)_2S(C_2H_4O)_7H$	0.1	22	25	5
Anionics				
$C_8H_{17}SO_4^-Na^+$	0.1	22	35	10
(br)*$C_8H_{17}SO_4^-Na^+$	0.1	22	30	4
$C_{8-10}H_{17-21}SO_4^-Na^+$	0.1	22	90	40
$C_{12}H_{25}SO_4^-Na^+$	0.1	22	155	135
$C_{12}H_{25}SO_4^-Na^+$ (0.1 M NaCl)	0.1	22	173	172
$C_{12}H_{25}SO_4^-NH_4^+$	0.1	22	176	161
$(C_{12}H_{25}SO_4^-)_2Mg^{2+}$	0.1	22	190	172
(br)*$C_{13}H_{27}SO_4^-Na^+$	0.1	22	98	30
$C_{12}H_{25}(OC_2H_4)SO_4^-Na^+$	0.1	22	175	160
$C_{12}H_{25}(OC_2H_4)SO_4^-Na^+$ (0.1 M NaCl)	0.1	22	157	155
$C_{12}H_{25}(OC_2H_4)_2SO_4^-Na^+$	0.1	22	170	157
$C_{12}H_{25}(OC_2H_4)_2SO_4^-Na^+$ (0.1 M NaCl)	0.1	22	157	156
$C_{12}H_{25}(OC_2H_4)_3SO4–Na+$	0.1	22	165	163
$C_{12}H_{25}(OC_2H_4)_3SO_4^-Na^+$ (0.1 M NaCl)	0.1	22	155	153
$C_{18}H_{35}SO_4^-Na^+$	0.1	22	156	145

(continued)

TABLE 2.7
(Continued)

Surfactants	C (%)	Temp 0°C	Foam (mm) Initial	Final (5 min)
Anionics (continued)				
$C_9H_{19}C_6H_4(OC_2H_4)_9SO_4^-NH_4^+$	0.1	22	185	180
$C_{12}H_{25}C_6H_4SO_3^-Na^+$	0.1	22	162	150
$C_{12}H_{25}C_6H_3(SO_3^-Na^+)OC_6H_4SO_3^-Na^+$	0.1	22	150	131
$C_4H_9CH(C_2H_5)O(O)CCH_2CH^-(SO_3^-Na^+)C(O)OCH_2CH(C_2H_5)C_4H_9$	0.1	22	168	154
$C_{12}H_{25}(OC_2H_4)_4OP(O)(OH)_2$ (pH 4)	0.1	22	170	170
$C_{12}H_{25}(OC_2H_4)_4OP(O)(O^-Na^+)_2$ (pH 7)	0.1	22	162	162
$C_{12}H_{25}(OC_2H_4)_4OP(O)(O^-Na^+)_2$ (pH 10)	0.1	22	151	150
$[C_6H_5(OC_2H_4)_6O]_{1,2}]P(O)(OH)_{1,2}$ (pH 7)	0.1	22	30	8
$[C_{8-10}H_{17-21}(OC_2H_4)_4O]_{1,2}P(O)(OH)_{1,2}$ (pH 4)	0.1	22	160	130
$[C_{8-10}H_{17-21}(OC_2H_4)_4O]_{1,2}P(O)(O^-Na^+)_{1,2}$ (pH 7)	0.1	22	155	140
$[C_{8-10}H_{17-21}(OC_2H_4)_4O]_{1,2}P(O)(O^-Na^+)_{1,2}$ (pH 10)	0.1	22	155	125
$[(br)^*C_{10}H_{21}(OC_2H_4)_4O]_{1,2}P(O)(O^-Na^+)_{1,2}$ (pH 7)	0.1	22	130	90
$[C_9H_{19}C_6H_4(OC_2H_4)_4O]_{1,2}P(O)(OH)_{1,2}$ (pH 7) (50 ppm Ca^{2+})	0.1	22	50	45
$[C_9H_{19}C_6H_4(OC_2H_4)_4O]_{1,2}P(O)(OH)_{1,2}$ (pH 7) (300 ppm Ca^{2+})	0.1	22	20	15
$[C_9H_{19}(C_6H_4)(OC_2H_4)_6O]_{1,2}P(O)(O^-Na^+)_{1,2}$ (pH 7)	0.1	22	133	120
$[C_9H_{19}(C_6H_4)(OC_2H_4)_6O]_{1,2}P(O)(O^-Na^+)_{1,2}$ (pH 7) (300 ppm Ca^{2+})	0.1	22	73	68
$[C_9H_{19}(C_6H_4)(OC_2H_4)_9O]_{1,2}P(O)(OH)_{1,2}$ (pH 3)	0.1	22	135	126
$[C_9H_{19}(C_6H_4)(OC_2H_4)_9O]_{1,2}P(O)(OH)_{1,2}$ (pH 7)	0.1	22	120	115
$[(C_9H_{19})_2C_6H_4(OC_2H_4)_7O]_{1,2}P(O)(OH)_{1,2}$ (pH 7)	0.1	22	15	5
$[C_{13}H_{27}C_6H_4(OC_2H_4)_3O]_{1,2}P(O)(O^-Na^+)_{1,2}$ (pH 7)	0.1	22	80	70
$[C_{13}H_{27}C_6H_4(OC_2H_4)_3O]_{1,2}P(O)(O^-Na^+)_{1,2}$ (pH 7) (300 ppm Ca^{2+})	0.1	22	30	25
$C_{11}H_{23}C(O)N(CH_3)CH_2CH_2SO_3^-Na^+$	0.05	25	93	80
$C_{11}H_{25}C(O)N(C_6H_{11})CH_2CH_2SO_3^-Na^+$	0.05	25	145	145
$C_{15}H_{31}C(O)N(CH_3)CH_2CH_2SO_3^-Na^+$	0.05	25	60	60
$C_{17}H_{33}C(O)N(CH_3)CH_2CH_2SO_3^-Na^+$	0.05	25	141	138
(Tall oil acyl)$C(O)N(CH_3)CH_2CH_2SO_3^-Na^+$	0.05	25	28	20
$(C_2F_5)_{3-8}CH_2CH_2SCH_2CH_2CO_2^-Li^+$	0.1	41	100	98
$(C_2F_5)_{3-8}CH_2CH_2SCH_2CH_2CO_2^-Li^+$	0.1	68	125	122
Zwitterionics				
$(br)^*C_8H_{17}N^+H(C_2H_4COOH)C_2H_4COO^-Na^+$ (pH 6)	0.1	22	4	1

(continued)

TABLE 2.7
(Continued)

		Foam (mm)		
Surfactants	C (%)	Temp 0°C	Initial	Final (5 min)
Zwitterionics (continued)				
$C_9H_{19}CONH(CH_2)_2N^+H(C_2H_4COOH)C_2H_4COO^-$ (pH 6)	0.1	22	22	18
$C_9H_{19}C(O)NH(CH_2)_2N^+H(CH_2CH_2OH)CH_2COO^-$ (pH 6)	0.1	25	17	13
$C_{11}H_{23}CONH(CH_2)_3N^+(CH_3)_2CH_2CH(OH)CH_2SO_3^-$ (pH 6)	0.1	22	153	153
$C_{11}H_{23}C(O)NH(CH_2)_2N^+H(C_2H_4OH)C_2H_4COO^-$ (pH 6)	0.1	22	165	165
$C_{11}H_{23}C(O)NH(CH_2)_2N^+H(CH_2CH_2OH)CH_2COO^-$ (pH 6)	0.1	25	151	151
$C_{11}H_{23}C(O)NH(CH_2)_3N^+(CH_3)_2CH_2COO^-$ (pH 6)	0.1	25	160	160
$C_{11}H_{23}C(O)NH(CH_2)_2N^+(CH_2CH_2OH)CH_2COOH$ (pH 6), with CH_2COO^- on N	0.1	25	156	156
(lauryl)$CONH(CH_2)_2N^+H(CH_2CH_2OH)CH_2COO^-$	0.1	22	175	175
$C_{12}H_{25}N^+H(C_2H_4COOH)C_2H_4COO^-Na^+$ (pH 6)	0.1	22	160	160
$C_{17}H_{35}C(O)NH(CH_2)_3N^+(CH_3)_2CH_2COO^-$ (pH 6)	0.1	25	152	150
$C_{12}H_{25}N^+(CH_3)_2O^-$ (pH 6)	0.1	22	160	160
$(C_2F_5)_{3-8}CH_2CH(OCOCH_3)CH_2N^+(CH_3)_2CH_2CO_2^-$ (pH 6)	0.1	41	220	220
$(C_2F_5)_{3-8}CH_2CH(OCOCH_3)CH_2N^+(CH_3)_2CH_2CO_2^-$ (pH 6)	0.1	68	220	220
Zwitterionic/anionic mixtures				
$C_{12}H_{25}N^+(CH_3)_2O^-$, $C_{12}H_{25}SO_4^-Na^+$	1.0	22	206	160
$C_{12}H_{25}N^+(CH_3)_2O^-$, $C_{12}H_{25}(OC_2H_4)_3SO_4^-Na^+$	1.0	22	204	181
$C_{11}H_{23}C(O)NH(CH_2)_2N^+H(C_2H_4OH)CH_2C(O)O^-$, $C_{12}H_{25}(OC_2H_4)SO_4^-Na^+$ (0.1 M NaCl)	0.1	22	170	170
$C_{11}H_{23}C(O)NH(CH_2)_2N^+H(C_2H_4OH)CH_2C(O)O^-$, $C_{12}H_{25}(OC_2H_4)_3SO_4^-Na^+$ (0.1 M NaCl)	0.1	22	168	166
$C_{11}H_{23}C(O)NH(CH_2)_3N^+(CH_3)_2CH_2C(O)O^-$, $C_{12}H_{25}(OC_2H_4)_3SO_4^-Na^+$ (0.1 M NaCl)	0.1	22	176	175
$C_{11}H_{23}C(O)NH(CH_2)_3N^+(CH_3)_2CH_2C(O)O^-$, $C_{12}H_{25}C_6H_4SO_3^-$ (0.1 M NaCl)	1.0	22	209	195

*Branched chain.

that adsorbs very rapidly at the surface of the solution. This method can make the time during which a surface tension difference exists between the thinning portions of the foam film and the surrounding film too short to bring in material to thicken the thinning portion. (Wetting agents often act as foam reducers by this mechanism.) Another method is to use a surfactant that adsorbs at the surface of the aqueous solution to form a loosely packed, poorly cohering film (e.g., A_{min} >> 100 Å^2/molecule) that has little stability.

Adsorption of Surfactants at the Surface of Nonaqueous Solutions

If adsorption of surfactants occurs at the surface of a nonaqueous solution, then we can expect wetting and foaming to occur, as previously discussed. However, in most nonaqueous systems, the two requirements for adsorption of the surfactant at the surface—(i) distortion of the solvent structure, and (ii) reduction of the dissimilarity between the gas phase and the solvent upon adsorption—are not met. Consequently, in most nonaqueous systems, adsorption of the surfactant at the surface does not occur to any significant extent and the solutions do not show significant surface tension reduction, foaming, or enhanced wetting.

Systems that do show adsorption at the surface include hydrocarbon chain surfactants in highly hydrogen-bonded solvents, such as ethylene glycol and glycerol, and perfluorocarbon chain surfactants in hydrocarbon solvents.

Adsorption onto Insoluble Solids or Liquids from Aqueous Solutions of Surfactants

As previously discussed, the adsorption of a surfactant from aqueous solution onto the surface of an insoluble liquid or solid can make that surface either more hydrophilic or more hydrophobic, depending on the nature of the surface sites on the liquid or solid.

Dispersion and Emulsification. When adsorption of the surfactant makes the surface of the liquid or solid (second phase) more hydrophilic, the interfacial tension between the aqueous solution and the second phase will be decreased and it will be easier to increase the area of that interface. If mechanical energy in the form of a grinder or mixer is now added to the system, it may be possible to break up the bulk second phase, to form an emulsion (in the case of a liquid second phase) or a dispersion (of the solid material) in the aqueous phase. The reduction of the interfacial tension between the water and the second phase makes this breakup of the bulk phase easier to occur, but it is not sufficient to give the resulting emulsion or dispersion significant stability. For the system to have significant stability, energy barriers surrounding the dispersed particles must be produced by the adsorption processes of sufficient height to prevent reaggregation of the particles.

These energy barriers may be electrical or steric in nature. Electrical barriers to the reaggregation of dispersed solid or liquid particles in aqueous media are produced by the adsorption onto the particles of ionic surfactants (Fig. 2.11a and b) in a quantity sufficient to give each particle a charge of the sign similar to that of the adsorbed surfactant. The greater the resulting electrical charge density on each particle, the greater the electrical barrier to reaggregation, since all the particles have a charge of the same sign and consequently repel each other. Steric barriers to reaggregation of dispersed particles in aqueous media are produced when adsorption of a nonionic surfactant (Fig. 2.11b) occurs with

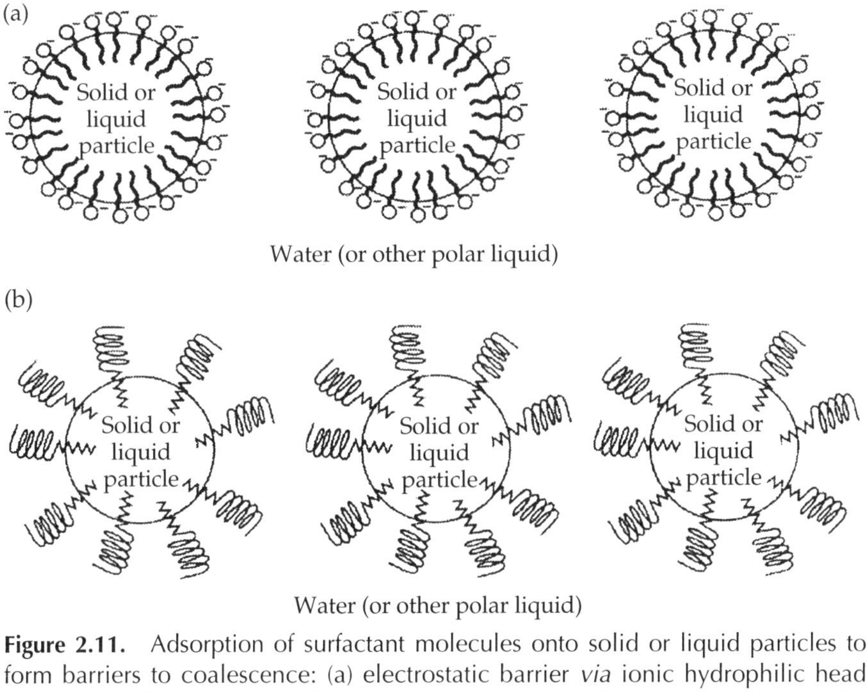

Figure 2.11. Adsorption of surfactant molecules onto solid or liquid particles to form barriers to coalescence: (a) electrostatic barrier *via* ionic hydrophilic head groups (anionic, in this case); (b) steric barriers *via* nonionic head groups (e.g., polyoxyethylene chains).

its hydrophilic head groups oriented toward the aqueous phase and these (hydrated) hydrophilic head groups are of sufficient bulk to constitute an energy barrier to the close approach of two dispersed particles. The adsorption of the surfactant in this fashion produces both enthalpic and entropic energy barriers to the reaggregation: enthalpic, in that close approach of the two particles (each covered with surfactant adsorbed in the same fashion) requires desolvation of the hydrophilic groups; entropic, in that this close approach also requires restriction of the free motion of these groups.

Flocculation and Demulsification. On the other hand, adsorption of the surfactant from the aqueous phase onto dispersed ionic or polar solid or liquid particles makes the particle surfaces more hydrophobic because the surfactant is adsorbed onto them with its hydrophilic head oriented toward the particle surface (Fig. 2.2a). In this case, the dispersion or emulsion will become more unstable than it was in the absence of the added surfactants, because (i) the particle surface/aqueous solution interfacial tension has been increased, thus increasing the fundamental instability of the system; and (ii) any electrical or steric barriers preventing aggregation of the dispersed particles will have been reduced by the surfactant adsorption. In some cases, the hydrophobic groups of the adsorbed surfactant molecules on two or more particles will aggregate, thus linking the particles to each other in order to minimize their contact with the aqueous phase.

Adhesion Promotion. Two surfaces adhere well when they are compatible with each other, i.e., similar in nature. As discussed earlier in this chapter, the adsorption of a surfactant at the surface of a solid or liquid can make it either more hydrophilic or more hydrophobic, depending on the orientation of the adsorbed surfactant film.

When we desire to promote the adhesion between two phases, surfactants should be selected which will, upon adsorption, make the surfaces of the two phases more similar in nature and attractive to each other. Thus, if we wish to cause a polar solid phase to adhere to a nonpolar hydrocarbon phase, a hydrocarbon chain surfactant that will adsorb onto the polar solid, with its hydrophilic head oriented toward the polar solid and its hydrocarbon group oriented away from it, should be used. This will allow the surface of the polar solid to become hydrocarbon-like in nature, and thus more prone to adhere to the hydrocarbon phase.

Adsorption onto Insoluble Solids and Liquids from Nonaqueous Solutions of Surfactants

Adsorption of surfactants can produce dispersions of insoluble second phases in nonaqueous liquids, as they do in aqueous media. If the dispersing liquid has a high dielectric constant, then both electrical and steric barriers are suitable for stabilizing the dispersion, as shown in Figure 2.11. In liquids of low dielectric constant, however, only steric barriers are suitable for stabilizing the dispersion, since electrical barriers are only of very short range, because of tight ion pair formation. Thus, magnetic oxide dispersions in hydrocarbon media, used for the production of magnetic tape, are stabilized by surfactant adsorption at the metal oxide/hydrocarbon interface with the hydrophilic heads oriented toward the polar oxide and the hydrophobic alkyl charges extending into the hydrocarbon phase (Fig. 2.3), thereby forming a steric barrier to the coalescence of the oxide particles.

References

1. Rosen, M.J., *Surfactants and Interfacial Phenomena*, 2nd edn., J Wiley, New York, 1989: (a) p. 65; (b) p. 68; (c) p. 84; (d) p. 69; (e) p. 43; (f) p. 249.
2. Rosen, M.J., *J. Am. Oil Chem. Soc. 51*:461 (1974).
3. Draves, C. Z., *Am. Dyestuff Rep. 28*:425 (1939).
4. Rosen, M.J., and X.Y. Hua, *J. Colloid Interface Sci. 139*:397 (1990).
5. Ross, J., and G.D. Miles, Am. Soc. for Testing Materials, Method D1173-53, Philadelphia, 1953; *Oil and Soap 18*:99 (1941).

CHAPTER 3

How Surfactants Change the Internal Properties of the Solution Phase and Related Performance Properties

As mentioned in Chapter 1, contact between the lyophobic group of the surfactant and the solvent can be minimized not only by adsorption at interfaces but also by aggregating the molecules of surfactant into spheres, cylinders, or sheets, called micelles, in which the lyophobic groups comprise the interior and the lyophilic groups comprise the exterior. Numerous studies have elucidated the conditions under which these aggregates form and the conditions under which these aggregates are transformed from one type of structure to another. The importance of this structural information in the utilization of surfactants is that it bears directly on various performance properties of the system. These performance properties include: (i) solubilization of normally solvent-insoluble material; (ii) hydrotropy, which is the inhibition of liquid crystal and crystal phase formation; and (iii) change in the viscosity of the solution phase. Since these properties are shown only by solutions that contain aggregated surfactant, and since the effects produced depend on the type of aggregated structure present, it is important to know when these structures first appear and the type of aggregate structure formed.

Micellization

The Critical Micelle Concentration

Very dilute surfactant solutions contain only monomeric surfactant molecules. When the concentration of surfactant in the solution is increased, a concentration range, called the **critical micelle concentration (cmc)**, is reached in which the individual surfactant molecules start to aggregate to form structures called micelles. If the solubility of the surfactant in the solvent at a particular temperature is low, then the cmc may not be reached at that temperature before the solubility limit is reached. In that case, if the solubility of the surfactant increases with an increase in temperature, then raising the temperature of the solution to the point at which the solubility equals the cmc of the surfactant at that temperature will produce micelles. That temperature is called the **Krafft point**. For micelles to be present in a solution, then, the solution must be above its Krafft point. Above its Krafft point, the solubility of the surfactant increases considerably.

A variety of methods are known for determining the value of the cmc, since almost all physical properties of the solution change when micelles are formed. Surface and interfacial tension, spectral changes with a variety of probes, and electrical conductivity are commonly used. The cmc of a surfactant depends on the chemical structure of the surfactant molecule and the environment in which it finds itself (the "molecular environment"), such as the temperature, pH, and electrolyte and organic additive content of the solution phase. Table 3.1 lists critical micelle concentration values of some commercial surfactants in aqueous media at 22°C.

In aqueous media free of salts and organic additives, the cmc (in molar units) decreases with an increase in the length of the hydrophobic group of the surfactant, by a factor of about 2 for a one carbon atom increase in the number of carbons in a straight alkyl chain of an ionic surfactant, and by a factor of about 3 for the same change in the case of nonionics or zwitterionics. For surfactants with similar hydrophobic groups, ionic surfactants have the highest cmc values, followed by zwitterionics, with nonionics having the smallest values. The presence of electrolytes in the solution phase can decrease the cmc values of ionics sharply. Ions of a charge opposite that of the surface-active ion (counterions), in the solution phase screen some of its charge, reducing the mutual repulsion of the similarly charged hydrophilic heads and making it easier for them to aggregate to form a micelle. As would be expected, divalent counterions depress the cmc more than monovalent ones. Thus, Ca^{2+} and Mg^{2+} depress the cmc of anionic surfactants to a greater degree than monovalent ions. For ions of the same charge, the smaller the radius of the hydrated ion, the larger its effect in depressing the cmc. For zwitterionics, the effect is smaller, and for nonionics even smaller.

Organic additives can increase or decrease the cmc values of all types of surfactants considerably, depending on the nature and concen-

TABLE 3.1
Critical Micelle Concentration (cmc) of Some Commercial Surfactants in Aqueous Medium at 22°C

Surfactants	CMC % by wt	CMC M
Nonionics		
$C_{8-10}H_{17-21}(C_2H_4O)_5(C_3H_7O)_5H$	0.002	4.33×10^{-5}
$C_{8-10}H_{17-21}(C_2H_4O)_6(C_3H_7O)_5H$	0.001	5.91×10^{-5}
$C_{9-11}H_{19-23}(OC_2H_4)_6OH$	0.025	8.92×10^{-4}
$C_{9-11}H_{19-23}(OC_2H_4)_8OH$	0.027	1.92×10^{-3}
$C_{10-12}H_{21-25}(OC_2H_4)_5OH$	0.0038	9.71×10^{-5}
$C_{10-12}H_{21-25}(OC_2H_4)_7OH$	0.0054	1.12×10^{-4}
(br)*$C_{10}H_{21}(OC_2H_4)_4OH$	0.026	7.80×10^{-4}
(br)*$C_{10}H_{21}(OC_2H_4)_6OH$	0.041	9.74×10^{-4}
$C_{10-12}H_{21-25}(OC_2H_4)_{10}Cl$	0.021	1.95×10^{-4}
(br)*$C_{10-12}H_{21-25}(OC_2H_4)_{10}Cl$	0.039	4.49×10^{-4}
$C_{10-12}H_{21-25}(OC_2H_4)_{14}Cl$	0.035	4.41×10^{-4}
$C_{12}H_{25}(OC_2H_4)_9OH$	0.0022	3.78×10^{-5}
$C_{12}H_{25}(OC_2H_4)_{12}OH$	0.0038	5.32×10^{-5}
$C_{12-13}H_{25-27}(OC_2H_4)_7OH$	0.0027	5.0×10^{-5}
$C_{12-14}H_{25-29}(OC_2H_4)_7OH$	0.0025	4.1×10^{-5}
$C_{12-14}H_{25-29}(OC_2H_4)_9OH$	0.0031	5.2×10^{-5}
$C_{12-14}H_{25-29}(OC_2H_4)_{12}OH$	0.0039	5.2×10^{-5}
$C_{12-14}H_{25-29}(OC_2H_4)_{15}OH$	0.0058	7.4×10^{-5}
$C_{12-15}H_{25-31}(OC_2H_4)_7OH$	0.0011	2.1×10^{-5}
$C_{12-15}H_{25-31}(OC_2H_4)_9OH$	0.0018	2.2×10^{-5}
$C_{14-15}H_{29-31}(OC_2H_4)_7OH$	0.0009	1.3×10^{-5}
$C_{14-15}H_{29-31}(OC_2H_4)_{13}OH$	0.0038	4.0×10^{-5}
$C_{16}H_{33}(OC_2H_4)_{15}OH$	0.00027	3.1×10^{-6}
$C_{18}H_{37}(OC_2H_4)_{20}OH$	0.00043	3.91×10^{-5}
p-t-$C_8H_{17}C_6H_4(OC_2H_4)_5OH$	0.0076	1.2×10^{-4}
p-t-$C_8H_{17}C_6H_4(OC_2H_4)_7OH$	0.0123	2.4×10^{-4}
p-t-$C_8H_{17}C_6H_4(OC_2H_4)_9OH$	0.0175	2.1×10^{-4}
p-t-$C_8H_{17}C_6H_4(OC_2H_4)_{12}OH$	0.0278	3.0×10^{-4}
p-t-$C_9H_{19}C_6H_4(OC_2H_4)_6OH$	0.0025	5.1×10^{-5}
p-t-$C_9H_{19}C_6H_4(OC_2H_4)_{9-10}OH$	0.0045	7.4×10^{-5}
p-t-$C_9H_{19}C_6H_4(OC_2H_4)_{11}OH$	0.0055	7.2×10^{-5}
p-t-$C_9H_{19}C_6H_4(OC_2H_4)_{15}OH$	0.0081	9.5×10^{-5}
p-t-$C_9H_{19}C_6H_4(OC_2H_4)_{20}OH$	0.0288	2.62×10^{-4}
$(C_9H_{19})_2C_6H_3(OC_2H_4)_9OH$	0.0031	4.22×10^{-5}
$(C_9H_{19})_2C_6H_3(OC_2H_4)_{15}OH$	0.0096	9.53×10^{-5}
$(C_9H_{19})_2C_6H_3(OC_2H_4)_{25}OH$	0.0325	2.25×10^{-4}
$C_{6-10}H_{13-21}(OC_3H_7)_4(OC_2H_4)_8(OC_3H_7)_{12}OH$	0.0035	2.45×10^{-5}

(continued)

TABLE 3.1
(Continued)

Surfactants	CMC	
	% by wt	M
Nonionics (continued)		
$(C_2F_5)_{3-8}CH_2CH_2O(C_2H_4O)_6H$	0.0014	9.2×10^{-5}
$C_9H_{19}C(CH_3)_2S(C_2H_4O)_7H$	0.0021	4.3×10^{-5}
Anionics		
$C_8H_{17}SO_4^-Na^+$	0.691	1.67×10^{-2}
(br)*$C_8H_{17}SO_4^-Na^+$	1.192	2.87×10^{-2}
$C_{12}H_{25}SO_4^-Na^+$	0.017	6.05×10^{-4}
$C_{12}H_{25}SO_4^-Na^+$ (0.1 M NaCl)	0.009	1.18×10^{-4}
$C_{12}H_{25}SO_4^-NH_4^+$	0.029	5.72×10^{-4}
$(C_{12}H_{25}SO_4^-)_2Mg^{2+}$	—	4.05×10^{-4}
$C_{12}H_{21}(OC_2H_4)SO_4^-Na^+$ (in 0.1 M NaCl)	0.0028	8.4×10^{-5}
$CC_{12}H_{25}(OC_2H_4)_2SO_4^-Na^+$ (in 0.1 M NaCl)	0.0023	6.2×10^{-5}
$C_{12}H_{21}(OC_2H_4)_3SO_4^-Na^+$ (in 0.1 M NaCl)	0.0026	4.7×10^{-5}
$C_9H_{19}C_6H_4(OC_2H_4)_4SO_4^-Na^+$	0.0079	4.6×10^{-4}
$C_9H_{19}C_6H_4(OC_2H_4)_4SO_4^-NH^+{}_4$	0.0078	6.6×10^{-4}
$C_9H_{19}C_6H_4(OC_2H_4)_9SO_4^-NH^+{}_4$	0.0654	9.2×10^{-4}
$C_{12}H_{25}C_6H_4SO_3^-NH^+{}_4$	0.029	8.5×10^{-4}
$[C_{12}H_{25}C_6H_3(SO_3^-NH^4)]OC_6H_4SO_3^-Na^+$	0.080	1.47×10^{-3}
$C_4H_9CH(C_2H_5)OOCCH_2CH(SO_3^-Na^+)COOCH_2CH$ $(C_2H_5)C_4H_9$	0.020	8.5×10^{-3}
$[C_8H_{17}(OC_2H_4)_3O)]_{1,2}P(O)(OH)_{1,2}$ (pH 3)	0.0157	1.81×10^{-4}
$[C_9H_{19}C_6H_4(OC_2H_4)_4O]_{1,2}P(O)(OH)_{1,2}$ (pH 3)	0.00054	8.05×10^{-6}
$[C_9H_{19}C_6H_4(OC_2H_4)_6O]_{1,2}P(O)(OH_{1,2}$ (pH 3)	0.00051	6.42×10^{-6}
$[C_9H_{19}C_6H_4(OC_2H_4)_9O]_{1,2}P(O)(OH)_{1,2}$ (pH 3)	0.00045	4.51×10^{-6}
$[C_{12}H_{25}(OC_2H_4)_4O]P(O)OH$ (pH 6)	0.0009	2.22×10^{-5}
(br)*$C_{13}H_{27}(OC_2H_4)_4O]_{1,2}P(O)(OH)_{1,2}$ (pH 3)	0.0019	3.02×10^{-5}
$C_{11}H_{23}C(O)N(CH_3)CH_2CH_2SO_3^-Na^+$	0.0175	5.1×10^{-4}
$C_{15}H_{31}C(O)N(CH_3)CH_2CH_2SO_3^-Na^+$	0.014	3.5×10^{-4}
$C_{15}H_{31}C(O)N(C_6H_{11})CH_2CH_2SO_3^-Na^+$	0.0153	3.3×10^{-4}
$C_{17}H_{33}C(O)N(CH_3)CH_2CH_2SO_3^-Na^+$	0.018	4.23×10^{-4}
(Tall Oil acyl)$C(O)N(CH_3)CH_2CH_2SO_3^-Na^+$	0.150	3.53×10^{-3}
$(C_2F_5)_{3-8}CH_2CH_2SCH_2CH_2CO_2^-Li^+$	0.012	1.51×10^{-4}
Zwitterionics		
(br)*$C_8H_{17}N^+H(CH_2CH_2COOH)CH_2CH_2COO^-$ (pH 6)	0.235	8.69×10^{-3}
$C_{12}H_{25}N^+H(CH_2CH_2COOH)CH_2CH_2COO^-Na^+$ (0.1 M NaCl) (pH 6)	0.002	5.76×10^{-5}
$C_9H_{19}C(O)NH(CH_2)_2N^+H(CH_2CH_2OH)CH_2COO^-$ (pH 6)	0.116	3.59×10^{-3}
$C_{11}H_{23}C(O)NH(CH_2)_2N^+H(CH_2CH_2OH)CH_2COO-$ (pH 6)	0.0069	2.04×10^{-4}

(continued)

TABLE 3.1
(Continued)

Surfactants	CMC % by wt	CMC M
Zwitterionics (continued)		
$C_{11}H_{23}C(O)NH(CH_2)_2N^+(C_2H_4OH)CH_2COOH$ (pH 6) with CH_2COO^- on N^+	0.0048	1.12×10^{-4}
$C_{11}H_{23}C(O)N(CH_2)_3N^+(CH_3)_2CH_2COO^-$ (pH 6)	0.0046	1.31×10^{-4}
$C_{17}H_{33}C(O)NH(CH_2)_3N^+(CH_3)_2CH_2COO^-$ (pH 6)	0.0035	9.31×10^{-5}
Cationics		
$C_{12}H_{25}N^+(CH_3)_3Br^-$	0.138	4.51×10^{-3}
$C_{12}H_{25}N^+(CH_3)_3Cl^-$	0.148	5.62×10^{-3}
$C_{12}H_{25}N^+(CH_2C_6H_5)(CH_3)_2Cl^-$	0.184	5.42×10^{-3}
$C_{16}H_{33}N^+(CH_3)_3Br^-$	0.018	6.02×10^{-4}
$C_{16}H_{33}N^+(CH_3)_3Cl^-$	0.027	6.72×10^{-4}
$C_{18}H_{35}N[(C_2H_4O)_{2.5}H]_2$	0.0012	2.50×10^{-5}
$C_{18}H_{35}N[(C_2H_4O)_{4.5}H]_2$	0.0029	9.50×10^{-5}
Zwitterionic/Anionic Salts		
$C_{11}H_{23}C(O)NH(CH_2)_2N^+H(C_2H_4OH)CH_2COO^-$, $C_{12}H_{25}OC_2H_4SO_4^-$	—	6.51×10^{-5}
$C_{11}H_{23}C(O)NH(CH_2)_3N^+(CH_3)_2CH_2COO^-$, $C_{12}H_{25}OC_2H_4SO_4^-$	—	4.55×10^{-5}
$C_{11}H_{23}C(O)NH(CH_2)_2N^+H(C_2H_4OH)CH_2COO^-$, $C_{12}H_{25}(OC_2H_4)_2SO_4^-$	—	3.81×10^{-5}
$C_{11}H_{23}C(O)NH(CH_2)_3N^+(CH_3)_2CH_2COO^-$, $C_{12}H_{25}(OC_2H_4)_2SO_4^-$	—	2.92×10^{-5}
$C_{11}H_{23}C(O)NH(CH_2)_2N^+H(C_2H_4OH)CH_2COO^-$, $C_{12}H_{25}(OC_2H_4)_3SO_4^-$	—	3.72×10^{-5}
$C_{11}H_{23}C(O)NH(CH_2)_3N^+(CH_3)_2CH_2COO^-$, $C_{12}H_{25}(OC_2H_4)_3SO_4^-$	—	2.56×10^{-5}

*Branched chain.

tration of the additive. This is particularly important in formulating commercial products, in which the presence of other ingredients in the formulation may change considerably the cmc of the surfactant from the value observed in the absence of the other ingredients. Polar compounds that are incorporated into the micelle depress the cmc (see Solubilization and Microemulsion Formation later in this chapter). Maximum depression of the cmc occurs when the polar additive has an alkyl group that is straight-chain and of approximately the same length as the surfactant. Compounds that are water structure breakers, such as urea, increase the cmc by increasing the hydration of the hydrophilic groups and decreasing the entropy change in the system resulting from the removal of the hydrophobic group from contact with water. Water structure formers, such as fructose, decrease the cmc.

Temperature increase affects the value of the cmc in aqueous media as a result of two phenomena that operate in opposite directions: (i) increased dehydration of the hydrophilic group, which favors micellization and decreases the value of the cmc; and (ii) increased disruption of the water structure, which disfavors micellization and increases the cmc value. For ionic surfactants in the range of 10–40°C, both effects appear to be small, and as a result, the value of the cmc changes only slightly. For polyoxyethylenated ("ethoxylated") nonionics, however, the first effect is much larger than the second, and the value of the cmc decreases sharply with the increase in temperature.

This dehydration of the polyoxyethylene group of polyoxyethylenated nonionics, caused by H-bond ruptures, results also in a decrease in the solubility of the surfactant in water, and it may separate out as a second phase. The temperature at which this separation occurs is known as the **cloud point** of the surfactant. The phenomenon is reversible, and a decrease in temperature causes the surfactant to redissolve clearly. At the cloud point, the surface-active properties of the surfactant change markedly. For example, these may cause the solution to lose its wetting and foaming abilities.

Micellar Shape and Aggregation Number

The micelles may be spherical, cylindrical, or lamellar (flat sheets) in shape, depending on the geometry of the monomeric surfactant units and their molecular environment.

A very useful parameter (1) for predicting the shape of the micelles in aqueous media from the dimensions of the surfactant molecule is

$$V_H/(l_c \cdot a_0) \qquad [3.1]$$

where V_H is the volume occupied by the hydrophobic group, l_c is (~80% of the fully extended length of the hydrophobic group, and a_0 is the cross-sectional area occupied by the hydrophilic group at the aqueous solution/micellar interface (approximately that occupied by the hydrophilic group at the aqueous solution/air interface). When the value of $V_H/(l_c \cdot a_0)$ is between 0 and one-third, then spherical micelles are formed in aqueous media; between one-third and one-half, cylindrical micelles are formed; and between one-half and 1, lamellar micelles are formed in aqueous media. When the value is >1, then reverse micelles (with the hydrophilic groups in the interior and the hydrophobic group in the exterior portion of the micelle) are formed in nonpolar media.

Whether the $V_H/(l_c \cdot a_0)$ ratio for a particular surfactant is greater or less than 1 can be determined by dissolving it in water (or an alkane), stirring the solution gently with an equal volume of alkane (or water), and observing in which phase the surfactant is at equilibrium. If it is in the aqueous phase, then $V_H/(l_c \cdot a_0)$ is less than 1; if it is in the alkane phase, it is more than 1. If the surfactant partitions into both phases, or if it separates out between them as a third phase, then the ratio is equal or close to 1.

Since the cross-sectional area occupied by the hydrophilic group at the interface often varies with the molecular environment of the surfactant (i.e., pH, ionic strength of the solution, temperature, presence of additives), it is possible to change the shape of the micelle by a change in the molecular environment. Thus, a_0 for ionic surfactants decreases with an increase in the ionic strength of the solution, due to greater screening of the electrical charge of their hydrophilic head groups by the increased concentration of counterions, effectively decreasing their repulsion of the similarly charged hydrophilic heads of adjacent surfactants, whereas a_0 for polyoxyethylenated surfactants decreases with an increase in the temperature of the solution, due to increased dehydration of the polyoxyethylene groups with the temperature increase. For ionic surfactants, a_0 also decreases with an increase in the concentration of the surfactant, since that also increases the concentration of counterions in the solution.

Since the ends of lamellar micelles have their lyophobic regions exposed to the solvent, they can curve up to form spherical structures called **vesicles**. Unilamellar vesicles consist of a single spherical layer of lamellar micelles surrounding a core of solvent, whereas multilamellar vesicles have concentric spheres of unilamellar vesicles, each separated from the other by a layer of solvent (Fig. 3.1)

The aggregation number of the surfactant micelle also depends on the geometric structure of the surfactant and the molecular environment. In aqueous media, an increase in the length, l_c, of the hydrophobic group results in an increase in the aggregation number, whereas an increase in the value of a_0 results in a decrease. An increase in the ionic strength of the solution causes an increase in the aggregation number of the micelles of ionic surfactants, because of the resulting reduction in the effective charge of their hydrophilic heads (mentioned above) and the consequent reduction in their mutual repulsion. A temperature

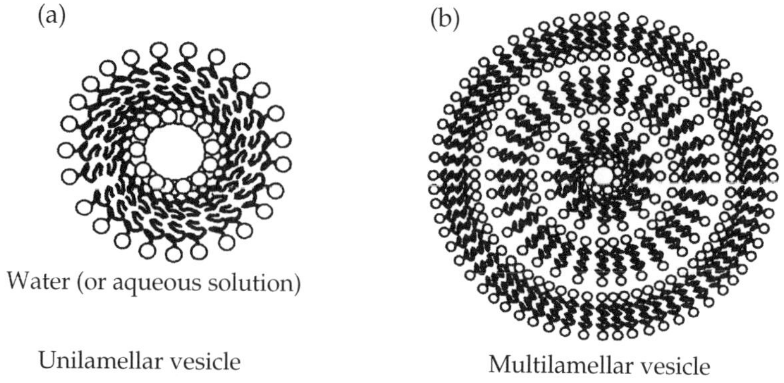

Figure 3.1. Vesicles: (a) unilamellar; (b) multilamellar.

increase causes a marked increase in the aggregation number of the micelles of polyoxyethylenated nonionics as a result of the increased dehydration of their polyoxyethylene groups.

Liquid Crystal Formation

As the concentration of micelles increases in a surfactant solution above its cmc, the micelles start to pack together in different arrangements that depend on the geometry of the individual micelles (Fig. 3.2). These packed arrangements are called **liquid crystals** and are intermediate between liquids and crystals in the arrangement and mobility of their components. They affect the viscosity of the solution greatly. Of particular importance in the utilization of surfactants are the hexagonal and lamellar liquid crystals that are formed by the packing together of cylindrical micelles and lamellar micelles, respectively. With an increase in

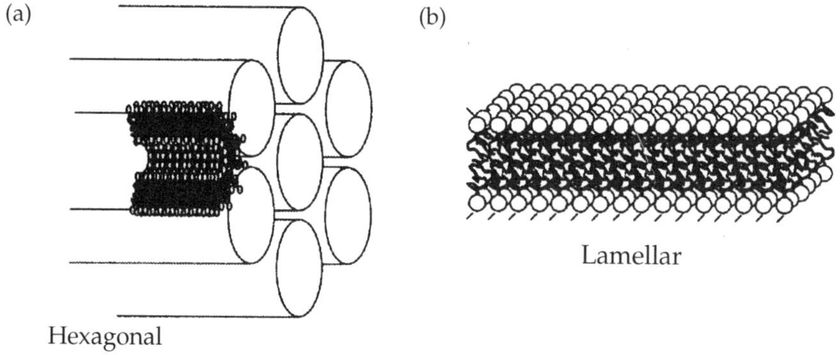

Figure 3.2. Hexagonal and lamellar liquid structures.

the concentration of the surfactant, a tendency often exists for surfactant micelles often to change from spherical to cylindrical, and then to lamellar. Consequently, hexagonal liquid crystals are usually encountered at lower surfactant concentrations than lamellar phases. Spherical micelles can pack together to form cubic liquid crystals, but these are usually encountered only at very high concentrations of surfactant, where they produce very high viscosity liquid crystal gels.

Hexagonal and lamellar liquid crystals are anisotropic structures and are detectable by their radiance when viewed under a polarizing microscope. They are also identifiable by ^{2}H NMR. Hexagonal liquid crystals can consist either of normal cylindrical micelles (having their hydrophobic groups oriented toward the interior of the cylinders and their hydrophilic groups on the surface) packed in a hexagonal arrangement, or of reverse micelles (with hydrophilic groups in the interior and hydrophobic groups on the surface) packed in a similar fashion. Under a polarizing microscope they appear as fan-like structures or with a variety of nongeometrical structures. Lamellar liquid crystals appear under the polarizing microscope as Maltese crosses or as oil streaks. Cubic liquid crystals are isotropic structures, as are spherical micelles, and are not observable with the polarizing microscope.

Relationship of Micellar Structure to Performance Properties

Solubilization and Microemulsion Formation

The presence of micelles in a solution makes possible the solubilization in it of materials (gas, liquid, or solid) that are normally insoluble in the solvent. The solubilized material dissolves in the micelles to form an optically clear, thermodynamically stable solution, and the system is therefore quite different from (macro)emulsions and other dispersions, which are opaque and thermodynamically unstable. (Of course, materials that are normally soluble in the solvent can also be dissolved or solubilized by micellar solutions.) The extent (moles solubilized/moles surfactant) to which a micellar solution can solubilize a particular material (solubilizate) depends on the location in the micelle at which the material is solubilized (Fig. 3.3). In aqueous media, short-chain polar materials are usually solubilized either on the surface of the micelle or in the outer region of the micelle close to the hydrophilic head groups, whereas nonpolar materials, such as aliphatic hydrocarbons, are solubilized deeply in the inner core of the micelle,

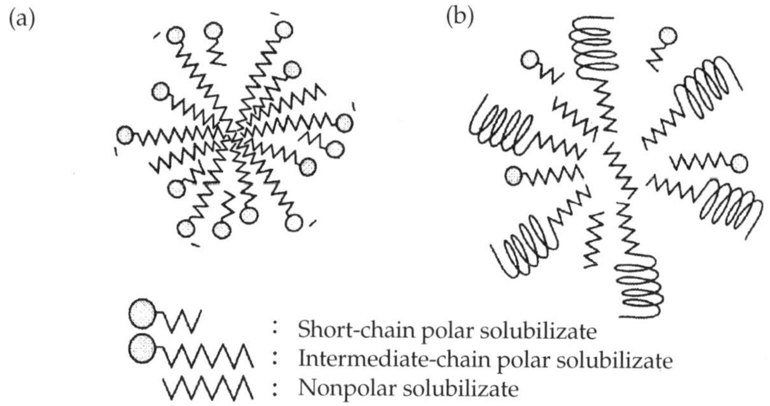

Figure 3.3. Location of solubilized material in (a) a spherical micelle of an ionic surfactant, and (b) a spherical micelle of a polyoxyethylenated nonionic surfactant.

between the hydrophobic groups. Polar materials of intermediate chain length will be solubilized in the micelle with their polar groups oriented toward the hydrophilic heads of the surfactant and their alkyl chains oriented toward the inner core of the micelle, their depth in the micelle increasing as the length of the alkyl chain increases.

Since the volume of a spherical micelle decreases with an increase in its depth into the micelle, materials that are solubilized deep in the micellar core are solubilized to a lesser extent than those solubilized closer to the surface of the micelle. For cylindrical micelles, the change in volume is much smaller with their increase in depth into the micelle, and for lamellar micelles essentially no change in volume occurs in going from the surface to the interior. As a result, nonpolar materials (that are solubilized deep in the interior) are solubilized to a greater extent in lamellar micelles than in spherical ones.

The extent of solubilization also depends on the aggregation number of the micelle: the larger the aggregation number, the greater the amount solubilized. The extent of solubilization also depends on the volume of the solubilizate, with small volume solubilizates solubilized to a greater extent than larger volume ones.

Since, in lamellar micelles, the volumes of the outer region of the micelle (where highly polar materials are solubilized) and the deep interior of the micelle (where nonpolar materials are solubilized) are approximately equal, this type of micelle should be able to solubilize approximately equal volumes of highly polar and nonpolar materials. Consequently, when $V_H/(l_c \cdot a_0)$ is approximately 1 (meaning that the cross-sectional area of the hydrophobic portion, a_h [= V_H/l_c], is approximately equal to that of the hydrophilic portion), a_0, of the surfactant molecule, the micellar solution will solubilize equal amounts of water and hydrocarbon. Since such solubilization can occur whether the original micellar solution is an aqueous one or a hydrocarbon one, the resulting system can consist either of normal micelles in aqueous solution swollen with hydrocarbon or of reverse micelles in hydrocarbon solution swollen with water. (In some cases, the system contains both types.) If the concentration of surfactant in its micellar form is sufficiently high, relatively large amounts of hydrocarbon and water can be present in the system and still form an optically clear mixture. These systems are called **microemulsions**. Microemulsions have found extensive use as cleaning agents, in emulsion polymerization, and in particular, have been intensively investigated for use in enhanced oil recovery from petroleum deposits.

Microemulsions can be formed by determining whether the $V_H/(l_c \cdot a_0)$ ratio of the surfactant(s) is greater or is less than 1 by the method described previously (see Micellar Shape and Aggregation Number) and then adjusting the ratio so that it is approximately 1 by changing the conditions. If the original value of the ratio is <1, then the ratio can be increased by: (i) increasing the length of the hydrophobic group of the surfactant; (ii) adding a polar cosurfactant of moderate chain length (C_6–C_{10}); (iii) adding an electrolyte, in the case of ionic surfactants; (iv) decreasing the polyoxyethylene chain length or increasing the temperature, in the case of polyoxyethylenated nonionics; or (v) decreasing the chain length of the alkane. If the original value of the ratio is >1, then the ratio can be decreased by: (i) decreasing the length of the hydrophobic group; or (ii) increasing the length of the polyoxyethylene chain, in polyoxyethylenated nonionics.

Hydrotropy

The presence of liquid crystals in surfactant solutions is often undesirable, since the close packing of their component micelles can limit the solubility of other solutes in the solution. Liquid crystal formation can be inhibited or eliminated by the use of surfactants or surfactant-like materials known as **hydrotropes**. They have short, bulky hydrophobic groups and large hydrophilic groups, preferably ionic, sometimes even two ionic groups in the molecule. They disrupt the parallel alignment of micellar structures in the hexagonal and lamellar liquid crystals and thereby inhibit their formation. Table 3.2 lists the hydrotropic abilities of some surfactants to dissolve

TABLE 3.2
Hydrotropic Activity of Some Commercial Surfactants for Coupling of Nonionic Surfactants

Basic Formulation[a,b]	A (%)	B (%)
Tetra potassium pyrophosphate[c]	10	10
$C_9H_{19}C_6H^4(OC_2H_4)_9OH$	5	10
Water	85	80

[a]2 phases, cloud point <20°C
[b]*Source:* Ref. 1.

Surfactants	Hydrotrope concentration (by wt) necessary for formulation compatibility 0–60°C	
Sodium xylene sulfonate	4.6	8.9
$C_4H_9(OC_2H_4)_4OP(O)(O^-Na^+)_{1,2}$	9.3	12.0
$[C_6H_5(OC_2H_4)_6O]_{1,2}P(O)(O^-Na^+)$ (mono/di 50/50)	3.9	5.4
$C_6H_5(OC_2H_4)_6OP(O)(O^-Na^+)$	2.8	4.6
$[C_6H_{13}(OC_2H_4)_4O]_{1,2}P(O)(O^-Na^+)$	2.0	3.9
$[C_8H_{17}C_6H_4(OC_2H_4)_6O]_{1,2}P(O)(O^-Na^+)$	2.1	4.5
$[C_9H_{19}C_6H_4(OCC_2H_4)_9O]_{1,2}P(O)(O^-Na^+)$	5.6	10.0
(br)*$C_8H_{17}(OC_2H_4)_4OP(O)(O^-Na^+)$	1.8	3.8
(br)*$C_8H_{17}SO_4^-Na^+$	2.6	4.1
$C_{10}H_{21}(OC_2H_4)_2SO_4^-Na^+$	3.7	6.2
$C_{12}H_{25}C_6H_4(SO_3^-Na^+)OC_6H_4SO_3^-Na^+$	4.0	9.8
$C_{12}H_{25}N(C_2H_4COO^-Na^+)_2$	1.8	3.0

Basic Formulation[c,d]	A (%)	B (%)
Sodium Hydroxide	5	10
$C_9H_{19}C_6H_4(OC_2H_4)_9OH$	5	5
Water	90	85

[c]2 phases, cloud point <20°C.
[d]*Source:* Ref. 2.
*Branched chain.

Surfactants	Hydrotrope concentration (by wt) necessary for formulation compatibility 0–60°C	
(br)*$C_8H_{17}N(C_2H_4COO^-Na^+)_2$[e] (pH 9)	0.5	1.8
$C_9H_{19}CONHC_2H_4N(C_2H_4OH)CH_2COO^-$	1.5	2.5
$C_9H_{19}CONHC_2H_4N(C_2H4OH)C_2H4COO^-$[e]	1.0	1.8
$C_{12}H_{25}N(C_2H4COO^-\ Na+)_2$[e]	0.4	1.4
$C_{12}H_{25}CONHC_2H_4N^+(C_2H_4OH)(C_2H_4COO^-)$ $C_2H_4COO^-Na^+$[e]	0.7	1.8

[e]All propionates are salt free.
*Branched chain.

a nonionic surfactant into aqueous solutions of electrolytes. Examples of hydrotropes are sodium m-xylene sulfonate, hexyl diphenyl ether disulfonate, and disodium alkylpolyoxyethylene phosphate.

Viscosity of Micellar Solutions

Aqueous solutions of spherical micelles at concentrations less than 20 times the cmc generally have viscosities not much different from that of water. As the surfactant concentration is increased, the micelles may become asymmetrical (cylindrical) and the solution may show a moderate increase in viscosity. When the concentration is increased further, the cylindrical micelles may pack together into a hexagonal arrangement to form, in some cases, long worm-like hexagonal liquid crystals, with a resulting great increase in the viscosity of the solution. A further increase in the surfactant concentration may result in the formation of lamellar liquid crystals from the hexagonal liquid crystals, with some reduction in the viscosity of the solution. In some cases, cubic liquid crystals may form during the transition from one liquid crystal form to another, with consequent changes in the viscosity of the system.

These viscosity changes are important, both because they may affect the ease of handling of the surfactant solutions for various applications, and also because they may affect the stability of dispersions such as foams and macroemulsions made with these surfactants. For example, the increased viscosity of the liquid film surrounding the air bubbles of a foam, as a result of the presence there of lamellar liquid crystals, reduces drainage of the liquid from the film, thus increasing stability of the foam. Lamellar liquid crystals surrounding the droplets can serve also to stabilize emulsions. By increasing the viscosity of the surfactant film surrounding the droplets and by increasing the viscosity of the solution phase, they decrease the rate of coalescence of the droplets.

References

1. Israelachvili, J.N., D.J. Mitchell, and B.W. Ninham, *J. Chem. Soc., Faraday Trans. I 72*:1525 (1976); *Biochiem. Biophys. Acta 470*:185 (1977).

CHAPTER 4

Chemical Structure and Microenviromental Effects on Fundamental Surfactant Properties and Related Performance Properties

The ability of surfactants to change the properties of interfaces or the properties of the solvent in which they are dissolved can be modified significantly by changes in the chemical structure of the surfactant and in the molecular environmental of the surfactant (pH, temperature, and electrolyte or other additive content of the solvent in which the surfactant is dissolved). Thus, the concentration of surfactant in the solution phase required to saturate the interface(s) (Chapter 2) or to form micelles in aqueous media (Chapter 3) decreases with an increase in the length of the alkyl chain of the hydrophobic group of the surfactant, whereas its ability to disperse particles in an aqueous medium increases with an increase in the ionic charge on its hydrophilic group. The pH of the solution affects the degree of ionization of surfactants containing carboxylate or nonquaternary ammonium groups, whereas increased electrolyte content decreases the effective distance to which the charge of ionic surfactants extends into the solution phase, and temperature affects such properties as solubility in the solvent, the viscosity of the solution phase, and the thermal agitation of the molecules both in the solution phase and at the interface(s).

Consequently, in order to understand why certain surfactants are used for a particular application and others are used in a different application, it is important to see how changes in the chemical structure and the molecular environment of the surfactant affect the properties relevant to its use.[1]

Solubility of Surfactants

Surfactants are used dissolved in a solvent, most commonly water or an aqueous solution, sometimes an aliphatic hydrocarbon mixture, rarely some other organic solvent. To be able to change either the properties of the interfaces or to change the internal properties of the solution phase, or both, the surfactant must be soluble in the latter but at the same time must distort its structure. To be active, then, the surfactant must be soluble but not too compatible with the solution phase.

In Aqueous Media

1. Solubility in aqueous media decreases with: (i) an increase in the length of the alkyl chain of the hydrophobic group (a phenyl group is equivalent to a 3–4 carbon alkyl chain); or (ii) an increase in the number of oxypropylene or oxybutylene groups in the molecule. The solubility of ionic surfactants in aqueous media decreases with: an increase in the ionic strength of the solution, particularly if a polyvalent counterion is present. The solubility of polyoxyethylenated and polyglycoside nonionic surfactants decreases with an increase in the temperature of the solution. The temperature at which they visibly come out of solution in water is known as the **cloud point**. At the cloud point, they may lose much of their surface or interfacial activity, but their solubilization capacity (see Solubilization later in this chapter) may increase considerably.
2. Solubility in aqueous media increases with: (i) an increase in the number of ionic groups in the molecule; (ii) an increase in the number of oxyethylene groups in the molecule; and (iii) an increase in branching or unsaturation in the hydrophobic group.

---

[1]It should be noted that the effects described below are generalizations to which there are exceptions.

3. Fatty acids that are insoluble in water may become soluble when the pH of the solution is raised above 7, due to formation of the corresponding fatty acid salt (soap). Long chain amines that are insoluble in water may become soluble when the pH is lowered below 7, due to formation of the corresponding long-chain ammonium salt.
4. Amphoteric surfactants (capable of both gaining and losing a proton) are generally more soluble above or below their isoelectric point (the pH at which the zwitterionic form is in equilibrium with equal amounts of the cationic and anionic forms of the surfactant). They have good solubility in the presence of electrolytes.
5. Phosphated esters of ethoxylated long-chain alcohols and alkylphenols ($\leq C_{12}$) have particularly good solubilities in the presence of acids, bases, and other electrolytes.

In Aliphatic Hydrocarbon Media

1. Solubility in aliphatic hydrocarbon media goes through a maximum with an increase in the length of the alkyl chain of a surfactant, and increases with an increase in the branching or unsaturation of the alkyl chain.
2. Solubility decreases with: (i) an increase in the number of ionic groups in the molecule; or (ii) an increase in the number of oxyethylene groups.
3. Solubility of anionic surfactants is increased by replacement of Na^+ as the counterion by Ca^{2+}, Ba^{2+}, ammonium, or alkylammonium.
4. Polyoxyethylenated nonionic surfactants that are soluble in aliphatic hydrocarbon media may become insoluble when the temperature is lowered in the presence of moisture or an aqueous phase. The temperature at which this occurs is known as the **"haze point."**
5. The free acids of phosphated polyoxyethylenated alcohols or polyoxyethylenated alkylphenols are frequently soluble in aliphatic hydrocarbons, as are long-chain fatty acid diesters of glycerol and the sodium salts of petroleum sulfonates of molecular weight >460.

Krafft Point

The Krafft point (Chapter 3, The Critical Micelle Concentration) increases with an increase in the length of the alkyl chain of the hydrophobic group and decreases with: (i) branching of the hydrophobic group; (ii) placement of oxyethylene or oxypropylene groups between the hydrophobic and hydrophilic groups in ionic surfactants; or (iii) replacement of a metal counterion in anionic surfactants by ammonium or alkylammonium counterions.

Alkyl sulfates have lower Krafft points than alkanesulfonates with the same number of carbon atoms in the alkyl chain and the same counterion.

Fluorocarbon chain surfactants have considerably higher Krafft points than hydrocarbon chain surfactants with the same number of carbon atoms and similar hydrophilic groups.

Krafft points for some commercial surfactants are listed in Table 4.1.

Cloud Point Formation

The cloud point of a nonionic polyoxyethylenated surfactant is increased by: (i) an increase in the oxyethylene content of the molecule; (ii) a decrease in the alkyl chain length—for straight-chain alcohols ethoxylated with six oxyethylene units, the cloud point is increased by 8–10°C when the alkyl chain is shortened by two carbon atoms; (iii) narrowing of the distribution of polyoxyethylene chain lengths; (iv) solubilization of nonpolar aliphatic hydrocarbons; (v) lowering of the pH of the solution with hydrochloric acid; or (vi) addition of some ionic surfactants to the solution. It can be increased or decreased by the addition of various electrolytes. Thus, NaCl, $NaNO_3$, Na_2SO_4, NaOH, and especially NaF decrease it, whereas NaI and NaSCN increase it.

The cloud points of alkyl polyglycoside nonionic surfactants increase with an increase in the average number of glycoside units ("degree of polymerization") in the molecule and, even more sharply than for alcohol ethoxylates, with a decrease in the alkyl chain length. Thus, $C_{12/14}$ alkyl polyglycosides have cloud points near room temperature; $C_{10/12}$ alkyl polyglycosides, above 100°C. All electrolytes, with the exception of NaOH, sharply reduce the cloud points of alkylpolyglycosides; NaOH increases it sharply.

Cloud points of some commercial nonionic solvents are listed in Table 4.2.

Electrical Effects

1. Adsorption of an ionic surfactant onto a substrate results in a decrease of a charge on that substrate that is opposite to that of the adsorbed ionic surfactant or an increase in a charge similar to that of the adsorbed ionic surfactant.

TABLE 4.1
Krafft Points of Some Commercial Anionic Surfactants

Surfactants	Krafft point (°C)
$C_8H_{17}SO_4^-Na^+$	<0
(br)*$C_8H_{17}SO_4^-Na^+$	<0
$C_{8-10}H_{17-21}SO_4^-Na^+$	<0
$C_{12}H_{25}SO_3^-Na^+$	22
$C_{12}H_{25}SO_4^-Na^+$	14
$C_{12}H_{25}SO_4^-Na^+$ (0.1 M NaCl)	20
$C_{12}H_{25}SO_4^-NH^+{}_4$	18
$(C_{12}H_{25}SO_4^-)_2Mg^{2+}$	16
$C_{12}H_{25}(OC_2H_4)SO_4^-Na^+$	10
$C_{12}H_{25}(OC_2H_4)SO_4^-Na^+$ (0.1 M NaCl)	18
$C_{12}H_{25}(OC_2H_4)_2SO_4^-Na^+$	8
$C_{12}H_{25}(OC_2H_4)_2SO_4^-Na^+$ (0.1 M NaCl)	14
$C_{12}H_{25}(OC_2H_4)_3SO_4^-Na^+$	<0
$C_{12}H_{25}(OC_2H_4)_3SO_4^-Na^+$ (0.1 M NaCl)	12
$C_{12}H_{25}C_6H_4SO_3^-Na^+$	16
(br)*$C_{12}H_{25}C_6H_4SO_3^-Na^+$	8
$C_9H_{19}C_6H_4(OC_2H_4)_4SO_4^-NH_4^+$	10
$C_9H_{19}C_6H_4(OC_2H_4)_9SO4_4^-NH_4^+$	<0
$C_9H_{19}C_6H_4(OC_2H_4)_{12}SO_4^-NH_4^+$	<0
(br)*$C_{13}H_{27}SO_4^-Na^+$	12
(br)*$C_{13}H_{27}OC_2H_4)_3OSO_4^-Na^+$	<0
$C_{16}H_{33}SO_4^-Na^+$	42
$C_{18}H_{35}SO_4^-Na^+$	20
$C_{12}H_{25}C_6H_3(SO_3Na^+)OC_6H_4SO_3^-Na^+$	<0
$C_4H_9CH(C_2H_5)OOCCH_2CH(SO_3^-Na^+)COOCH_2CH(C_2H5)C_4H_9$	18
$C_{11}H_{23}C(O)N(CH_3)CH_2CH_2SO_3^-Na^+$	5
$C_{11}H_{25}C(O)N(C_6H_{11})CH_2CH_2SO_3^-Na^+$	<0
$C_{15}H_{31}C(O)N(CH_3)CH_2CH_2SO_3^-Na^+$	10
$C_{17}H_{33}C(O)N(CH_3)CH_2CH_2SO_3^-Na^+$	<0
(Tall oil acyl) $C(O)N(CH_3)CH_2CH_2SO_3^-Na^+$	<0

*Branched chain.

2. The addition of an inert electrolyte (A^+B^-) to an aqueous solution of an ionic surfactant decreases the effective charge of the ionic hydrophilic group of the surfactant. When the electrolyte contains polyvalent ions of a charge opposite that of the ionic hydrophilic group of the surfactant, the effective charge of the latter is reduced even more.

TABLE 4.2
Cloud Points of Some Commercial Nonionic Surfactants

Surfactants	Solvent (other than water)	Cloud point (°C)
$C_{6-10}H_{13-21}(OC_3H_7)_4(OC_2H_4)_8(OC_3H_7)_{12}OH$		18–20
$C_{8-10}H_{17-21}(C_2H_4O)_4(C_3H_7O)_5H$		25–27
$C_{8-10}H_{17-21}(OC_2H_4)_5(OC_3H_7O)_5OH$		25–27
$C_{8-10}H_{17-21}(C_2H_4O)_6(C_3H_7O)_4H$		30–32
$C_{8-10}H_{17-21}(OC_2H_4)_6(OC_3H_7)_7OH$		12–14
$C_{9-11}H_{19-23}(OC_2H_4)_6OH$		52
$C_{9-11}H_{19-23}(OC_2H_4)_6OH$	5%, TKPP[a]	36
$C_{9-11}H_{19-23}(OC_2H_4)_6OH$	5%, STPP[a]	34
$C_{9-11}H_{19-23}(OC_2H_4)_6OH$	5%, Na_2SO_4	37
$C_{9-11}H_{19-23}(OC_2H_4)_6OH$	5%, Na_2SiO_3	26
$C_{9-11}H_{19-23}(OC_2H_4)_6OH$	5%, NaOH	19
$C_{9-11}H_{19-23}(OC_2H_4)_6OH$	5%, Na_2CO_3	14
$C_{9-11}H_{19-23}(OC_2H_4)_6OH$	5%, H_2SO_4	51
$C_{9-11}H_{19-23}(OC_2H_4)_6OH$	5%, HCl	61
$C_{9-11}H_{19-23}(OC_2H_4)_8OH$		80
$C_{9-11}H_{19-23}(OC_2H_4)_8OH$	5%, TKPP	58
$C_{9-11}H_{19-23}(OC_2H_4)_8OH$	5%, STPP	58
$C_{9-11}H_{19-23}(OC_2H_4)_8OH$	5%, Na_2SO_4	52
$C_{9-11}H_{19-23}(OC_2H_4)_8OH$	5%, Na_2SiO_3	47
$C_{9-11}H_{19-23}(OC_2H_4)_8OH$	5%, NaOH	41
$C_{9-11}H_{19-23}(OC_2H_4)_8OH$	5%, Na_2CO_3	31
$C_{9-11}H_{19-23}(OC_2H_4)_8OH$	5%, H_2SO_4	81
$C_{9-11}H_{19-23}(OC_2H_4)_8OH$	5%, HCl	89
$C_{10}H_{21}(OC_2H_4)_5(OC_3H_7)_5OH$		22–25
$C_{10-12}H_{21-25}(OC_2H_4)_7(OC_3H_7)_7OH$		30–35
(br)*$C_{10}H_{21}(OC_2H_4)_4OH$		<20
(br)*$C_{10}H_{21}(OC_2H_4)_6OH$		40–43
(br)*$C_{10}H_{21}(OC_2H_4)_{10}Cl$		28–32
(br)*$C_{10}H_{21}(OC_2H_4)_{14}Cl$		42–44
$C_{11}H_{23}(OC_2H_4)_5OH$		<20
$C_{11}H_{23}(OC_2H_4)_7OH$		45–48
$C_{11}H_{23}(OC_2H_4)_{10}Cl$		30–32
$C_{11}H_{23}(OC_2H_4)_{14}Cl$		40–42
$C_{12}H_{15}(OC_2H_4)_7OH$		49–51
$C_{12}H_{25}(OC_2H_4)_7OH$	5%, TKPP[a]	37
$C_{12}H_{25}(OC_2H_4)_7OH$	5%, STPP[a]	35
$C_{12}H_{25}(OC_2H_4)_7OH$	5%, Na_2SO_4	30
$C_{12}H_{25}(OC_2H_4)_7OH$	5%, Na_2SiO_3	29
$C_{12}H_{25}(OC_2H_4)_7OH$	5%, NaOH	22

(*continued*)

***TABLE 4.2**
(Continued)

Surfactants	Solvent (other than water)	Cloud point (°C)
$C_{12}H_{25}(OC_2H_4)_7OH$	5%, Na_2CO_3	15
$C_{12}H_{25}(OC_2H_4)_7OH$	5%, H_2SO_4	49
$C_{12}H_{25}(OC_2H_4)_7OH$	5%, HCl	58
$C_{12}H_{25}(OC_2H_4)_9OH$		74–76
$C_{12}H_{25}(OC_2H_4)_{12}OH$		90–95
$C_{12-13}H_{25-27}(OC_2H_4)_6OH$		42–46
$C_{12-13}H_{25-27}(OC_2H_4)_7OH$		44–46
(br)*$C_{12-14}H_{25-29}(OC_2H_4)_7OH$[b]		37–39
(br)*$C_{12-14}H_{25-29}(OC_2H_4)_9OH$[b]		60–62
(br)*$C_{12-14}H_{25-29}(OC_2H_4)_{12}OH$[b]		88–92
(br)*$C_{12-14}H_{25-29}(OC_2H_4)_{15}OH$[b]		>100
$C_{12-15}H_{25-31}(OC_2H_4)_7OH$		47–53
$C_{12-15}H_{25-31}(OC_2H_4)_9OH$		72–76
$C_{12-15}H_{25-31}(OC_2H_4)_{12}OH$		86–88
(br)*$C_{13}H_{27}(OC_2H_4)_3OH$		<20
(br)*$C_{13}H_{27}(OC_2H_4)_9OH$		56–60
(br)*$C_{13}H_{27}(OC_2H_5)_{15}OH$		>100
$C_{14-15}H_{29-31}(OC_2H_4)_7OH$		44–46
$C_{14-15}H_{29-31}(OC_2H_4)_{13}OH$		>100
$C_{16-20}H_{33-42}(OC_2H_4)_{10}(OC_3H_7)_3OH$		54
$C_{16-20}H_{33-42}(OC_2H_4)_{10}(OC_3H_7)_4OH$		50
$C_{16-20}H_{33-42}(OC_2H_4)_{20}(OC_3H_7)_7OH$		55
$C_{18}H_{33}(OC_2H_4)_5OH$		<20
$C_{18}H_{33}(OC_2H_4)_{20}OH$		>100
p-t-$C_8H_{17}C_6H_4(OC_2H_4)_5OH$		<20
p-t-$C_8H_{17}C_6H_4(OC_2H_4)_7OH$		<20
p-t-$C_8H_{17}C_6H_4(OC_2H_4)_9OH$		64–66
p-t-$C_8H_{17}C_6H_6(OC_2H_4)_9OH$	5%,TKPP[a]	49
p-t-$C_8H_{17}C_6H_6(OC_2H_4)_9OH$	5%, STPP[a]	48
p-t-$C_8H_{17}C_6H_6(OC_2H_4)_9OH$	5%, Na_2SO_4	41
p-t-$C_8H_{17}C_6H_6(OC_2H_4)_9OH$	5%, Na_2SiO_3	39
p-t-$C_8H_{17}C_6H_6(OC_2H_4)_9OH$	5%, NaOH	30
p-t-$C_8H_{17}C_6H_6(OC_2H_4)_9OH$	5%, Na_2CO_3	19
p-t-$C_8H_{17}C_6H_6(OC_2H_4)_9OH$	5%, H_2SO_4	67
p-t-$C_8H_{17}C_6H_6(OC_2H_4)_9OH$	5%, HCl	76
p-t-$C_8H_{17}C_6H_4(OC_2H_4)_{12}OH$		88–92
p-t-$C_8H_{17}C_6H_4(OC_2H_4)_{15}OH$		>100
p-t-$C_9H_{19}C_6H_4(OC_2H_4)_4CH_2C_6H_5$		26–28
p-t-$C_9H_{19}C_6H_4(OC_2H_4)_6OH$		<20

(continued)

TABLE 4.2
(Continued)

Surfactants	Solvent (other than water)	Cloud point (°C)
p-t-$C_9H_{19}C_6H_4(OC_2H_4)_9OH$		52–56
p-t-$C_9H_{19}C_6H_4(OC_2H_4)_9OH$	5%, TKPP[a]	38
p-t-$C_9H_{19}C_6H_4(OC_2H_4)_9OH$	5%, STPP[a]	37
p-t-$C_9H_{19}C_6H_4(OC_2H_4)_9OH$	5%, Na_2SO_4	30
p-t-$C_9H_{19}C_6H_4(OC_2H_4)_9OH$	5%, Na_2SiO_3	29
p-t-$C_9H_{19}C_6H_4(OC_2H_4)_9OH$	5%, NaOH	19
p-t-$C_9H_{19}C_6H_4(OC_2H_4)_9OH$	5%, Na_2CO_3	8
p-t-$C_9H_{19}C_6H_4(OC_2H_4)_9OH$	5%, H_2SO_4	54
p-t-$C_9H_{19}C_6H_4(OC_2H_4)_9OH$	5%, HCl	64
p-t-$C_9H_{19}C_6H_4(OC_2H_4)_{10}OH$		60–63
p-t-$C_9H_{19}C_6H_4(OC_2H_4)_{12}OH$		>100
p-t-$C_9H_{19}C_6H_4(OC_2H_4)_{12}OH$	10%, NaCl	50–54
p-t-$C_9H_{19}C_6H_4(OC_2H_4)_{15}OH$	10%, NaCl	64–66
p-t-$C_9H_{19}C_6H_4(OC_2H_4)_{20}OH$	10%, NaCl	68–70
$C_{12}H_{25}C_6H_4(OC_2H_4)_9OH$		36–40
$C_{12}H_{25}C_6H_4(OC_2H_4)_{11}OH$		62–66
$C_{18}H_{35}C(O)O(C_2H_4O)_8(OC_3H_7)_4OH$		<20
$C_{18}H_{37}C(O)O(C_2H_4O)_{23}(OC_3H_7)_{22}OH$		41
$(C_9H_{19})_2C_6H_3(OC_2H_4)_{15}OH$		48–52
$(C_9H_{19})_2C_6H_3(OC_2H_4)_{25}OH$		>100
Tristyrylphenol ethoxylates (16 OE)		53–57
Tristyrylphenol ethoxylates (20 OE)		81–85
Tristyrylphenol ethoxylates (25 OE)		>100
Nopol* $(C_3H_7O)_3$ $(C_2H_4O)_5H$		18–20
Nopol $(C_3H_7O)_5$ $(C_2H_4O)_6H$		40–45
Nopol $(C_3H_7O)_3$ $(C_2H_4O)_7H$		60–65
Nopol $(C_3H_7O)_4$ $(C_2H_4O)_5H$		12–15
Nopol $(C_3H_7O)_4$ $(C_2H_4O)_7H$		40–43
Nopol $(C_3H_7O)_4$ $(C_2H_4O)_9H$		58–60
Nopol $(C_3H_7O)_4$ $(C_2H_4O)_{11}H$		70-72
Nopol $(C_3H_7O)_5$ $(C_2H_4O)_5H$		<20
Nopol $(C_3H_7O)_5$ $(C_2H_4O)_7H$		40–42
Nopol $(C_3H_7O)_5$ $(C_2H_4O)_9H$		48–50
Nopol $(C_3H_7O)_5$ $(C_2H_4O)_{11}H$		63–65
Nopol $(C_3H_7O)_5$ $(C_2H_4O)_{13}H$		72–75
$HO(C_2H_4O)_2(OC_3H_7)_{16}(OC_2H_4)_2OH$		37
$HO(C_2H_4O)_{11}(OC_3H_7)_{16}(OC_2H_4)_{11}OH$		77
$HO(C_2H_4O)5(OC_3H_7)_{21}(OC_2H_4)_5OH$		37
$HO(C_2H_4O)_7(OC_3H_7)_{21}(OC_2H_4)_7OH$		42
$HO(C_2H_4O)_{11}(OC_3H_7)_{21}(OC_2H_4)_{11}OH$		67

(continued)

TABLE 4.2
(*Continued*)

Surfactants	Solvent (other than water)	Cloud point (°C)
$HO(C_2H_4O)_3(OC_3H_7)_{30}(OC_2H_4)_3OH$		24
$HO(C_2H_4O)_7(OC_3H_7)_{30}(OC_2H_4)_7OH$		28
$HO(C_2H_4O)_8(OC_3H_7)_{30}(OC_2H_4)_8OH$		32
$HO(C_2H_4O)_{13}(OC_3H_7)_{30}(OC_2H_4)_{30}OH$		61
$HO(C_2H_4O)_8(OC_3H_7)_{35}(OC_2H_4)_8OH$		25
$HO(C_2H_4O)_{24}(OC_3H_7)_{35}(OC_2H_4)_{24}OH$		82
$HO(C_2H_4O)_6(OC_3H_7)_{39}(OC_2H_4)_6OH$		20
$HO(C_2H_4O)_{22}(OC_3H_7)_{39}(OC_2H_4)_{22}OH$		74
$HO(C_2H_4O)_{27}(OC_3H_7)_{39}(OC_2H_4)_{27}OH$		85
$HO(C_9H_{19})_2C_6H_3(OC_2H_4)_9OH$		<20
$C_{18}H_{37}C(O)O(C_2H_4O)_{80}(OC_3H_7)_{38}OH$		58
Tall oil $(OC_2H_4)_{16}OH$		73
Castor oil $(OC_2H_4)_5OH$		<20
Castor oil $(OC_2H_4)_{16}OH$		61
Castor oil $(OC_2H_4)_{30}OH$		71
Castor oil $(OC_2H_4)_{36}OH$		74
Castor oil $(OC_2H_4)_{40}OH$		78
Castor oil (hyd) $(OC_2H_4)_{25}OH$		74
Castor oil (hyd) $(OC_2H_4)_{20}OH$		93
$C_{18}H_{37}N[(C_2H_4O)_{2.5}]_2H$		<0
$C_{18}H_{37}N[(C_2H_4O)_{7.5}]_2H$		30–31
$C_9H_{19}(CH_3)_2S(C_2H_4O)_7H$		16–18
$C_9H_{19}(CH_3)_2S(C_2H_4O)_8H$		26–28
$C_9H_{19}(CH_3)_2S(C_2H_4O)_{10}H$		50–52
$[(CH_3)_3SiO]_2Si(CH_3)(CH_2)_3(OC_2H_4)_{7.5}OCH_3$		<20

*Nopol

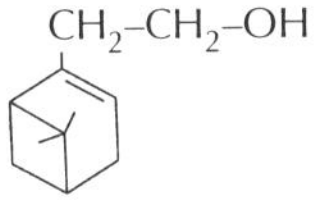

[a]Abbreviations: TKPP; tetrapotassium pyrophosphate; STPP, sodium tripolyphosphate.
[b]Secondary alcohols.
*Branched chain.

Packing at Interfaces

1. The packing of a surfactant adsorbed from an aqueous solution onto the surface of a hydrophobic substrate is generally determined by the cross-sectional area of the hydrophilic group, rather than of the hydrophobic group, unless the latter is very highly branched or unsaturated.

2. Molecules having a single, long hydrophobic group and two hydrophilic groups separated from each other by a short hydrophobic group are generally more loosely packed than molecules with a single hydrophobic and a single hydrophilic group.
3. When the molecule has two long hydrophobic groups and two hydrophilic groups separated by a short-chain linkage ("gemini" or "dimeric" surfactants, Chapter 5), the packing of the hydrophobic groups is closer than in surfactants having a single, similar hydrophobic and single, similar hydrophilic group.
4. Packing increases slightly with an increase in the length of a straight alkyl chain hydrophobic group from ten to sixteen carbon atoms and decreases with branching or unsaturation in the alkyl chain, or with the introduction of a cyclic structure.
5. Packing of an ionic surfactant is increased considerably by: (i) an increase in the ionic strength of the aqueous phase; (ii) the presence of polyvalent counterions in the aqueous phase; and (iii) the presence of an oppositely charged surfactant.
6. The packing of a polyoxyethylenated alcohol or alkylphenol at the air/aqueous solution interface can be decreased considerably by replacing the terminal hydroxyl of the polyoxyethylene chain with a short hydrophobic group.

Reduction of Surface Tension

1. In surfactants of a given charge type, the efficiency of surface tension reduction in aqueous media increases with an increase in the length of the hydrophobic group. That is, less of the longer chain compound is needed to reduce the surface tension to some desired value.
2. The rate of reduction of the surface tension of an aqueous solution is increased by: (i) branching in the hydrophobic group; (ii) an increase in the area of the surfactant molecule at the surface; or (iii) reduction of the molecular weight of the surfactant at constant HLB[2].
3. Hydrocarbon-chain surfactants can reduce the equilibrium surface tension of water to about 25 dyn/cm (mN/m); fluorocarbon-chain surfactants to about 15 dyn/cm (mN/m); and silicone-chain (polydi-

[2]The HLB (hydrophilic–lipophilic balance is a number indicative of the ratio of polar to nonpolar character of the surfactant. For ethoxylated nonionics, it is weight percentage of oxyethylene in the molecule divided by 5.

methylsiloxane) surfactants to about 20 dyn/cm (mN/m), at a minimum.

4. The introduction of branching into the hydrocarbon chain of a surfactant decreases the value to which it can reduce the equilibrium surface tension of water.
5. The addition of an inert electrolyte (A^+B^-) or of an oppositely charged surfactant decreases the value to which an ionic surfactant can reduce the equilibrium surface tension of water.

Dispersion of Solids in Liquid Media

1. The dispersing ability of ionic surfactants for solids in aqueous media generally increases with an increase in the number of ionic groups in the surfactant molecule.
2. For the dispersion of polar or ionic solids in aqueous media, the hydrophobic group of the surfactant should contain polar groups, such as ether linkages, or polarizable groups, such as aromatic rings, capable of adsorbing onto the solid.
3. Dispersing agents for polar or ionic solids in hydrocarbon media generally have polarizable lyophobic groups and long alkyl chains that can extend into the solvent. They can be monomeric or polymeric.

Emulsification

1. Mixtures of polyoxyethylenated nonionic surfactants containing both short polyoxyethylene chains (low HLB)[2] and long polyoxyethylene chains (high HLB)[2] produce better emulsions than surfactants with a narrow range of polyoxyethylene content.
2. The polarity (HLB) of the emulsifier mixture required to yield a good emulsion increases as the polarity of the oil to be emulsified in an oil-in-water (O/W) emulsion increases.
3. Good emulsifying agents should have limited solubility in both the oil and water phases of the system. Too great a solubility in either phase decreases the efficiency and effectiveness of the emulsifying agent.
4. Tighter packing of the hydrophobic groups at the oil/water interface (see Packing at Interfaces earlier in this chapter) increases the stability of the (macro)emulsion.

Foaming

1. Foam height goes through a maximum with an increase in the length of the alkyl chain of the hydrophobic group and is decreased considerably by branching in that group.
2. Foaming is decreased considerably by any structural factor that increases the cross-sectional area of the surfactant at the surface (see Packing at Interfaces earlier in this chapter), e.g., branching in the hydrophobic group, an increase in the number of oxyethylene units in the hydrophilic group of nonionics, or "capping" of the terminal hydroxyl of the polyoxyethylene group with a less water-soluble group.
3. Foaming of polyoxyethylenated nonionics decreases sharply at or above their cloud points.
4. Ionic surfactants are better foamers than nonionics, presumably because their ionic head groups increase the stability of the foam through electrostatic repulsion between the two sides of the foam lamella.
5. Foaming reaches its maximum height in the vicinity of the cmc of the surfactant.

Solubilization

1. The extent of solubilization (moles of solubilizate/moles of surfactant) of polar solubilizates in spherical micelles in aqueous media is much greater than for nonpolar solubilizates; nonpolar material is solubilized to a greater extent than polar solubilizates in reverse spherical micelles in aliphatic hydrocarbon solutions.
2. The addition of an inert electrolyte (A^+B^-) to an aqueous ionic micellar solution, with the consequent change in micellar shape from spherical to cylindrical to lamellar, increases the extent of solubilization of nonpolar solubilizates and decreases that of polar solubilizates.
3. An increase in temperature increases the solubilization extent of both nonpolar and polar materials in aqueous micellar solutions of ionic surfactant. For nonionic polyoxyethylenated surfactants, as the temperature approaches the cloud point, a considerable decrease occurs in the extent of solubilization of polar material and a considerable increase occurs in that of nonpolar material.

Wetting by Aqueous Solutions

1. Surfactants of relatively small molecular weight, those having a branched hydrophobic group and a centrally located hydrophilic group that makes the molecule just barely soluble in water, produce the fastest wetting.
2. As the temperature is raised, the optimum chain length for fastest wetting increases when the solubility of the surfactant increases with the temperature increase.
3. For polyoxyethylenated nonionic surfactants, the fastest wetting is shown by those whose cloud points are just above the use temperature.
4. Polyoxyethylenated alcohols and mercaptans show faster wetting than corresponding esters of fatty acids.
5. The area to which a surfactant-containing aqueous solution can spread over a substrate depends on the surfactant's ability to depress both the surface tension of the wetting liquid and the substrate/liquid interfacial tension. Therefore, trisiloxane-based surfactants that can depress the surface tension of water to 21 dyn/cm produce larger spreading areas of water on hydrocarbon surfaces than hydrocarbon-chain surfactants, which can depress the surface tension to no lower than 25 dyn/cm. On the other hand, fluorocarbon-chain surfactants that can depress the surface tension of water to 15 dyn/cm do not produce as large spreading areas of water on hydrocarbon surfaces as trisiloxane-based surfactants. This is because the lack of affinity of fluorocarbons for hydrocarbon surfaces does not depress the wetting liquid/hydrocarbon interfacial tension as much as does a trisiloxane-chain surfactant.

CHAPTER 5

Enhancing the Performance of Surfactants

Synergism

Nearly all commercial surfactants, with very few exceptions, are mixtures. They are mixtures because the hydrophobe raw materials used in their manufacture are almost always mixtures of homologs and, in addition, various isomers of the nominal surfactant structure are formed in their synthesis. Thus, in the preparation of the common surfactant, sodium linear alkylbenzenesulfonate (LAS), although the alkyl chain length may be given in the manufacturer's literature as C_{12} (dodecyl), other chain lengths (e.g., C_{11}, C_{13}) are invariably present in the commercial product. In addition, the benzenesulfonate groups are attached at different positions (C_2, C_3, C_4, etc.) in the alkyl chain. Finally, dialkyltetralinsulfonates are also present, along with the alkylbenzenesulfonates. The term "surfactant" then, when used to designate a commercial product, is understood to mean a mixture of this type.

In mixtures of surfactants that are related to each other and, especially, that are of the same charge type, the overall performance properties are determined in large part by the most highly surface active species that is present in the mixtures to a significant extent (see Chapter 4, Reduction of Surface Tension, for structural effects on surface activity and performance properties). On the other hand, if a surfactant (whether an individual material or a mixture of the same charge type) is mixed with a surfactant of a different charge type (e.g., an anionic with a cationic, or a cationic with a zwitterionic), the possibility exists that the resulting mixture will show better properties than those of the individual surfactants. When the mixture shows better properties than either component of the mixture by itself at the same concentration in the solution, the system is said to exhibit **synergism**.

Mixing different charge types of surfactants that may exhibit synergism is an important method of enhancing performance properties. When synergism is present: (i) less mixed surfactant can be used to obtain the same level of performance, with the consequent economic and/or environmental benefit, and (ii) the level of performance of the synergistic mixture (foaming, emulsifying ability) may be greater than obtainable with either component by itself.

Investigation of the phenomenon of synergism in mixtures of surfactants reveals that it depends on an attractive molecular interaction occurring between the two different surfactants (1, and references therein). This is, however, a necessary but not sufficient requirement. The existence of synergism in the system also depends on the values of the desired property in the two different surfactants. For example, when reducing the critical micelle concentration (or increasing the adsorption efficiency at an interface), the greater the difference between the cmc values (or C_{20} values) of the two surfactants, the stronger must be the attractive molecular interaction between them for synergism to exist (see below).

Calculating the Molecular Interaction (β) Parameter Between Surfactant Pairs

The nature and strength of the molecular interaction between two surfactants is measured by what is called the "beta (β) parameter." The β parameter is related to the free energy yielded when the two surfactants are mixed. Attractive interactions yield negative β values; repulsive interactions, positive values. When no interaction occurs, the system shows ideal mixing, and the value of the β parameter is zero. The greater the value of the β parameter, the stronger the interaction, either attractive or repulsive. The value of the parameter also depends on the nature of the interface at which the two surfactants interact; values are generally different at different interfaces. Values of the β parameters have been measured for interaction at the aqueous solution/air interface (β^σ), the aqueous solution/micelle interface (β^m), the aqueous solution/hydrocarbon interface (β^σ_{LL}), and the aqueous solution/hydrophobic solid interface (β_{LS}). The value of β^σ or β^m is obtained from surface tension–concentration data; β^σ_{LL}, from liquid/liquid interfacial tension–concentration data; β_{LS}, from a combination of surface tension and contact angle–concentration data or surface tension–concentration and adsorption isotherm data.

Figure 5.1 illustrates how values of β^σ or β^m are calculated from surface tension (γ)–concentration (C) data. γ is plotted against log (or ln) C for each of the surfactant components of the mixture by itself and also for at least one mixture of them at some fixed mole fraction, α_1, on a surfactant-only basis, i.e., moles of surfactant 1/(total moles of surfactant 1 plus surfactant 2). C_1°, C_2°, and C_{12} are the concentrations of surfactants 1, 2, and their mixture, respectively, required to obtain the same surface tension in their respective solutions. The value of γ selected should be close to the minimum common value. cmc_1, cmc_2, and cmc_{12} are the cmc values of surfactants 1, 2, and their mixture, respectively.

The experimental values C_1°, C_1°, and C_{12}, and β are used in Equation 5.1,

$$\frac{X_1{}^2 \ln(\alpha_1 \ln[(1-\alpha_1)\, C_{12}/X_1 C_1^\circ)}{(1-X_1)^2 \ln[(1-\alpha_1)C_{12}/(1-X_1)C_2^\circ]} = 1 \qquad [5.1]$$

which is solved iteratively for the value of X_1, the mole fraction of surfactant 1 in the total surfactant at the aqueous solution/air interface. The value of X_1, obtained in this fashion, is used in Equation 5.2,

$$\beta^\sigma = \frac{\ln(\gamma_1 C_{12}/X_1 C_1^\circ)}{(1-X_1)^2} \qquad [5.2]$$

to calculate β^σ.

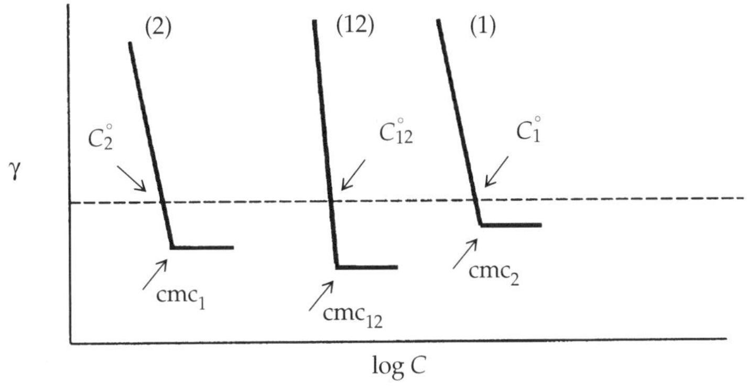

Figure 5.1. Calculation of the values of β^σ or β^m from plots of surface tension (γ)– concentration data: (1), surfactant 1; (2), surfactant 2; (12), mixture of surfactants 1 and 2 at a fixed value of α_1, the mole fraction of surfactant 1 in the total surfactant of the mixture (on a surfactant-only basis).

For calculating β_m, the interaction parameter at the aqueous solution/micelle interface, the values of the critical micelle concentrations, cmc_1, cmc_2, and cmc_{12}, for surfactant 1, surfactant 2, and their mixture, respectively, are substituted in Equation 5.3,

$$\frac{(X_1^m)^2 \ln(\alpha_1 \, cmc_{12}/X_1^m cmc_1)}{(1 - X_1^m)^2 \ln[(1 - \alpha_1)cmc_{12}/(1 - X_1^m)cmc_2]} = 1 \qquad [5.3]$$

which is solved iteratively to calculate X_1^m, the mole fraction of surfactant 1 in the total surfactant at the aqueous solution/micelle interface. The value of X_1^m is then substituted in Equation 5.4,

$$\beta^m = \frac{\ln(\gamma_1 cmc_{12}/X_1^m cmc_1)}{(1 - X_1^m)^2} \qquad [5.4]$$

to calculate β^m.

For calculating β_{LL}^{σ}, the interaction at the aqueous solution/hydrocarbon interface, interfacial tension–log (or ln) C plots are used in analogous fashion.

To calculate β_{LS}^{σ} at the aqueous solution/solid interface of a nonporous, nonpolar hydrocarbon substrate, contact angles, θ, on a planar smooth surface of this solid, are measured and $\cos\theta$ (see Eq. 2.6) for values below the cmc of the solution phase is plotted vs. log (or ln) C (Fig. 5.2). Values of C_1°, C_2°, and C_{12} are determined at a common value of $\gamma\cos\theta$ close to the largest common value. When contact angles cannot be obtained on a smooth, planar, nonporous surface of the solid, β_{LS} values can be obtained from the adsorption isotherms for solutions of surfactant 1, 2, and at least one mixture of them at a fixed value of α_1, on the finely divided solid. By integrating the area under the plot of Γ vs. log (or ln) C (Fig. 2.5) at various values of log (or ln) C below the respective cmc's of the surfactants, the values of Π, the amount of interfacial tension reduction at the solid/liquid interface, are determined. These are plotted (Fig. 5.3) and values of C_1°, C_2°, and C_{12}, at the same values of Π are selected for substitution into Equations 5.1 and 5.2 for the evaluation of β_{LS}. A Π value should be selected that is close to the largest common value.

Values of β parameters for mixtures of some commercial surfactants are listed in Table 5.1.

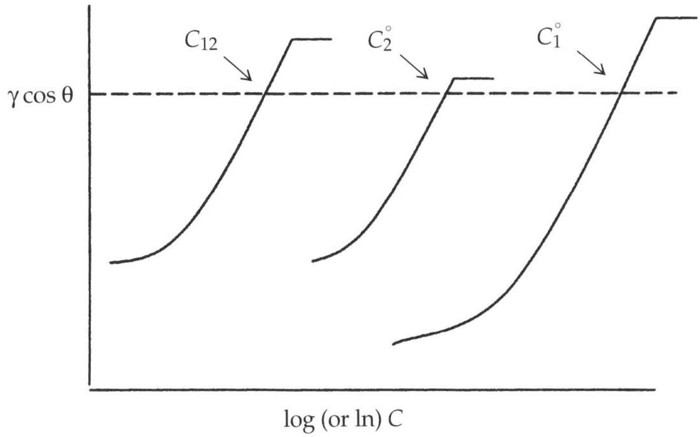

Figure 5.2. Calculation of β_{LS}^{σ} from surface tension and contact angle–concentration data below the cmc of the solution phase: (1) surfactant 1; (2) surfactant 2; (12) mixture of surfactants 1 and 2 at a fixed value of α_1, the mole fraction of surfactant 1 in the total surfactant of the mixture (on a surfactant-only basis).

From β parameter values calculated in this fashion, the attraction between two different surfactants has been found to be dominated by electrostatic forces between the hydrophilic groups; van der Waals forces between the hydrophobic groups also play a role, but a relatively minor one. For mixtures containing an anionic surfactant as one component, the attractive interaction between the two different charge types decreases in this order: anionic/cationic > anionic/zwitterionic capable of accepting a proton > anionic/ethoxylated nonionic > anionic/other nonionic surfactant.

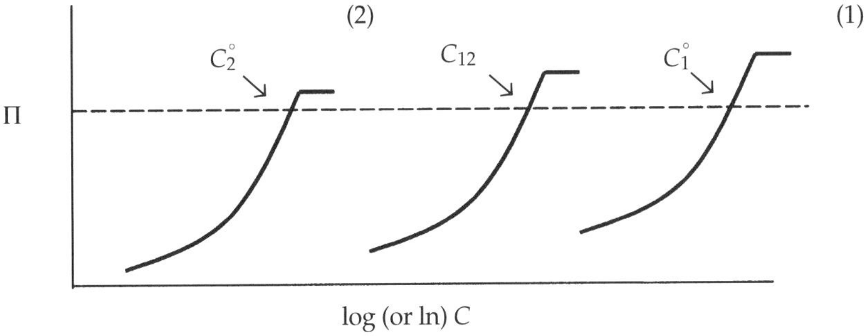

Figure 5.3. Evaluation of C_1°, C_2°, and C_{12}, for the determination of β_{LS}^{σ} from adsorption isotherm data below the cmc of the solution phase: (1), surfactant 1; (2), surfactant 2; (12), mixture of surfactants 1 and 2 at a fixed value of α_1.

TABLE 5.1
β^σ and β^m Values of Some Commercial Surfactant Mixtures, pH = 6

System	β^σ	β^m
$C_{12}AmNH^+(E_t)CH_2COO^-/C_{12}OESO_4^-Na^{+a}$	−5.2	−3.3
$C_{12}AmN^+(E_t)(CH_2COOH)(CH_2COO^-)/C_{12}OESO_4^-$	−9.4	−3.5
$C_{12}AmpN^+(CH_3)_2CH_2COO^-/C_{12}OESO_4^{-a}$	−6.6	−1.2
$C_{12}AmNH(E_t)CH_2COO^-/C_{12}(OE)_2SO_4^-Na^+$	−6.5	−2.1
$C_{12}AmN^+(E_t)(CH_2COO^-Na^+)_2/C_{12}(OE)_2SO_4^-Na^+$	−9.0	−2.2
$C_{12}AmpN^+(CH_3)_2CH_2COO^-)/C_{12}(OE)_2SO_4^-$	−4.6	−2.6
$C_{12}AmN(E_t)CH_2COO^-/C_{12}(OE)_3SO_4^-Na^+$	−7.3	−2.6
$C_{12}Am^+N(E_t)(CH_2COO^-Na^+)_2/C_{12}(OE)_3SO_4^-$	−15.7	−2.6
$C_{12}AmpN^+(CH_3)_2CH_2COO^-/C_{12}(OE)_3SO_4^-$	−5.7	−3.9
$C_{12}Am\ N^+(E_t)(CH_2COO^-Na^+)_2$/DTAB	−17.7	−12.6
$C_{12}(OE)_3SO_4^-Na^+$/DTAB	−16.7	−6.90
SDBS/DTAB[a]	−21.3	−13.7

[a]Abbreviations: Am, $-CONH\ CH_2\ CH_2^-$; Amp, $-CONH\ CH_2\ CH_2\ CH_2^-$; DTAB, ditallow dimethyl ammonium bromide; SDBS, sodium salt of dodecylbenzene sulfonic acid.

For mixtures containing a cationic surfactant as one component, the order of decreasing attractive interaction is: cationic/anionic > cationic/zwitterionic capable of losing a proton > cationic/nonionic surfactant.

Structural factors that increase the attractive interaction between two different types of surfactants are: an increase in the length of the alkyl chains of the two surfactants and equal lengths of the two alkyl chains of the two surfactants (especially at interfaces). On the other hand, moving the hydrophilic group to a more central position in the molecule decreases the strength of its interaction with an oppositely charged surfactant. An increase in the ionic strength of the solution generally decreases the attractive interaction of ionic surfactants, since it reduces the effective charge of the ionic hydrophilic groups. A decrease in the pH of the solution increases the attractive interaction between anionic surfactants and zwitterionics capable of accepting a proton, whereas an increase in the pH decreases it. In analogous fashion, an increase in the pH of the solution increases the interaction between cationic surfactants and zwitterionics capable of losing a proton, whereas a decrease in the pH decreases it.

Requirements for Synergism

As mentioned above, synergism between two surfactants depends not only on mutual attraction between them, but also on the values of the relevant property of the individual components of the mixture. The exact requirements for synergism, resulting from either mixed monolayer formation by the two surfactants at an interface or from mixed micelle formation by them in the aqueous phase, have been determined (1, and references therein). Synergism in these fundamental properties of surfactants is the basis for synergism in such performance properties as foaming, wetting, emulsification, and detergency (as shown later in this chapter).

Mixed monolayer formation at an interface can produce synergism in surface (or interfacial) tension reduction *efficiency*, or synergism in surface (or interfacial) tension reduction *effectiveness*, or both. Synergism in surface (or interfacial) tension reduction *efficiency* exists when a given surface (or interfacial) tension value can be obtained at a lower surfactant concentration of the mixture than of either of the two components of the mixture by itself. Surface (or interfacial) tension reduction *effectiveness* exists when the mixture can reach a lower surface (or interfacial) tension than can either of the two components of the mixture by themselves.

Mixed micelle formation in the aqueous phase can result in synergism in micelle formation. When this occurs, the critical micelle concentration of the mixture is smaller than that of either component of the mixture by itself, i.e., the mixture has a greater tendency to form micelles than either component alone.

The requirements for synergism in surface tension reduction efficiency are:

1. β^{σ} must be negative, and [5.5]
2. $|\beta^{\sigma}|$ must be greater than $|\ln(C_1^{\circ}/C_2^{\circ})|$. [5.6]

The requirements for synergism in mixed micelle formation are:

1. β^{m} must be negative, and [5.7]
2. $|\beta^{m}|$ must be greater than $|\ln(\text{cmc}_1/\text{cmc}_2)|$. [5.8]

The requirements for synergism in surface tension reduction effectiveness are:

1. $\beta^{\sigma} - \beta^{m}$ must be negative, and [5.9]
2. $|\beta^{\sigma} - \beta^{m}|$ must be greater than $|\ln C^{H}/C^{L}|$. [5.10]

The C_1°, C_2°, cmc_1, and cmc_2 values are obtained as shown in Figure 5.1. C^H and C^L are obtained as shown in Figure 5.4.

For synergism in interfacial tension reduction efficiency or effectiveness at the aqueous solution/nonpolar liquid hydrocarbon or at the aqueous solution/nonpolar hydrocarbon solid interface, β_{LL}, or β_{LS}, respectively, values are substituted for β^{σ} in conditions 5.5, 5.6, 5.9, and 5.10.

In synergism in surface (or interfacial) tension reduction efficiency, or for mixed micelle formation to exist in a mixture, relationships 5.5–5.8 indicate that: (i) the interaction between them at the relevant interface must be attractive (5.5 and 5.7); and (ii) the greater the difference between the tendency of the two individual surfactants either to reduce surface tension ($|\ln(C_1^{\circ}/C_2^{\circ})|$), or to form micelles ($|\ln(cmc_1/cmc_2)|$), the stronger must be the interaction between them for synergism to exist. Therefore, to obtain maximum synergism in either of these two respects, the two surfactants chosen should have, if possible, strong electrostatic interaction between their hydrophilic head groups and should have similar surface active tendencies (similar C_{20} or cmc values, respectively).

For synergism in surface (or interfacial) tension reduction effectiveness to exist, relationships 5.9 and 5.10 indicate that: (i) the attractive interaction between the two surfactants at the interface must be stronger than their attractive interaction in the micelle; and (ii) if the two surfactants have the same γ (or γ_1) value at their respective cmc's, irrespective of their cmc values, then $\ln(C^H/C^L)$ will be zero and any negative value of $(\beta^{\sigma} - \beta^{m})$ will produce synergism in this respect (condition 5.10 and Fig. 5.3).

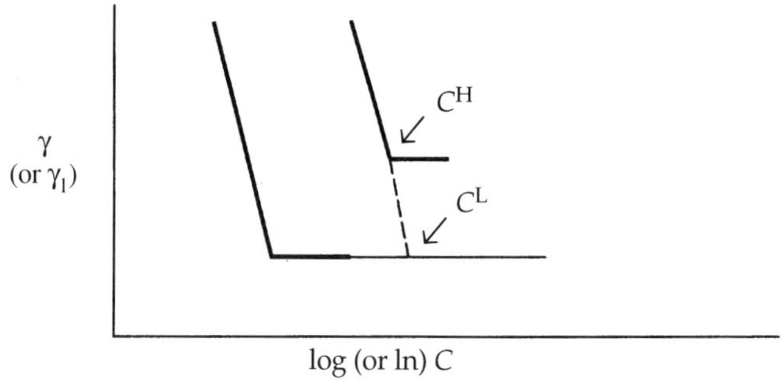

Figure 5.4. Extensions of the linear portion of the γ (or γ_1) plot of log (or ln) C below the cmc, for the surfactant having the higher γ (or γ_1) value at its cmc(C^H), in order to evaluate C^L.

Also, the steeper the slope of the γ (or γ_1) value–ln *C* plot for the surfactant with the larger γ (or γ_1) at its cmc, the closer the value of $\ln(C^H/C^L)$ will approach zero, making synergy in this respect more probable. Therefore, to obtain synergy of this type (i.e., to produce a mixture that will have a lower surface (or interfacial) tension value than obtainable with either surfactant component), the surfactant pair should be selected to have hydrophobic groups having approximately the same alkyl chain length, since this usually produces β^σ values more negative than β^m values. (This satisfies condition 5.9.) In addition, they should be chosen to have surface (or interfacial) tension values at their respective cmc's that are close to each other, or the slope of the γ (or γ_1)–log (or ln) *C* plot of the surfactant having the higher γ (or γ_1) value at its cmc should be steep (in order to satisfy condition 5.10).

In summary, then, synergism of any type is more likely to exist when the two surfactants used for the mixture show the strongest molecular attraction for each other: an anionic mixed with a cationic, or with a zwitterionic capable of being protonated; a cationic mixed with an anionic, or with a zwitterionic capable of being deprotonated. For obtaining synergism in surface (or interfacial) tension reduction efficiency, or for obtaining synergism in mixed micelle formation, the two surfactants chosen should have values of the desired property close to each other (C_{20} values in the case of surface or interfacial tension reduction efficiency, and cmc values in the case of mixed micelle formation). For obtaining synergism in surface (or interfacial) tension reduction effectiveness, the two surfactants chosen should have alkyl chains close in length to each other; their surface (or interfacial) tension values at their respective cmc's should be close to each other; or the surfactant having the higher tension value at its cmc should have a γ (or γ_1)–log (or ln) *C* plot with a steep slope. It should be noted that mixtures of anionic surfactants and nonionic ethoxylated surfactants with about six or more oxyethylene units in the nonionic surfactant molecule show β^σ values less negative than their β^m values, and consequently do not meet condition 5.9 for synergism in surface tension reduction effectiveness.

If the two surfactants to be mixed have approximately equal values of the desired property (C_{20}, cmc, $_{cmc}$, or $_{1cmc}$), then equimolar amounts of the two should be used to obtain maximum synergy; if they do not, then the one with the property closer to the desired value should be used in larger molar amount than the other.

Mixtures of two surfactants can also show antagonism (negative synergism), in which case the mixture shows less surface activity than the individual components (e.g., less efficient or less effective surface tension reduction or a higher cmc value). In these cases, the sign of the β parameter is positive, indicating a repulsive interaction between the two surfactants. Mixtures of fluorocarbon-chain and hydrocarbon-chain surfactants of the same charge type show repulsive interaction, as do mixtures of long-chain soaps with long-chain alkanesulfonates or with long-chain alkylbenzenesulfonates. In some cases, such repulsive interactions are desirable. For example, the addition of soap to alkylbenzenesulfonate formulations reduces the amount of foam produced by the latter. This is desirable in low-foaming cleaning products. The action is due to the repulsive interaction between the soap and alkylbenzenesulfonate molecules, resulting in an increase in the surface tension of the mixture and a reduction in foaming. The conditions for antagonism (negative synergism) are completely analogous to those for synergism except that the values of β^σ and β^m must be positive (conditions 5.5, 5.7, and 5.9).

Synergism in surface tension reduction effectiveness has been correlated with increased wetting (Table 5.2, Figs. 5.5 and 5.6) and with increased initial foaming by aqueous surfactant mixtures (Table 5.3), whereas synergism in interfacial tension reduction effectiveness in oil/water (O/W) systems has been correlated with increased dishwashing ability and detergency.

TABLE 5.2
Synergism in Surface Tension Reduction and Wetting Time for Some Zwitterionic and Anionic Surfactants and their Mixtures at pH = 6, T = 25°C, NaCl = 0.1 M

Surfactant mixture	γ (cmc) (expt dyn/cm)	Draves wetting (0.1 wt%/sec)
$C_{12}AmN^+H(E_t)CH_2COO^-$[a]	26.5	34.0
$C_{12}AmN^+(E_t)(CH_2COO)_2^-$	28.3	60.0
$C_{12}AmpN^+(CH_3)_2CH_2COO^-$[a]	32.1	65.0
$C_{12}(OE)_2SO_4^-Na^+$	31.2	19.2
$C_{12}(OE)_3SO_4^-Na^+$	34.0	40.0
$C_{12}AmN^+H(E_t)CH_2COO^-/C_{12}(OE)_2SO_4^-Na^+$	26.4	16.0
$C_{12}AmN^+H(E_t)(CH_2COO^-)_2/C_{12}(OE)_2SO_4^-Na^+$	27.9	16.3
$C_{12}AmpN^+H(CH_3)_2CH_2COO^-)/C_{12}(OE)_2SO_4^-Na^+$	28.8	13.8
$C_{12}AmN^+H(E_t)CH_2COO^-/C_{12}(OE)_3SO_4^-Na^+$	25.8	18.5
$C_{12}AmN^+(E_t)(CH_2COO)_2/C_{12}(OE)_3SO_4^-Na^+$	27.5	24.0
$C_{12}AmpN^+(CH_3)_2CH_2COO^-/C_{12}(OE)_3SO_4^-Na^+$	29.4	20.0

[a]Abbreviations: Am, $-CONHCH_2CH_2^-$; Amp, $-CONHCH_2CH_2CH_2^-$.

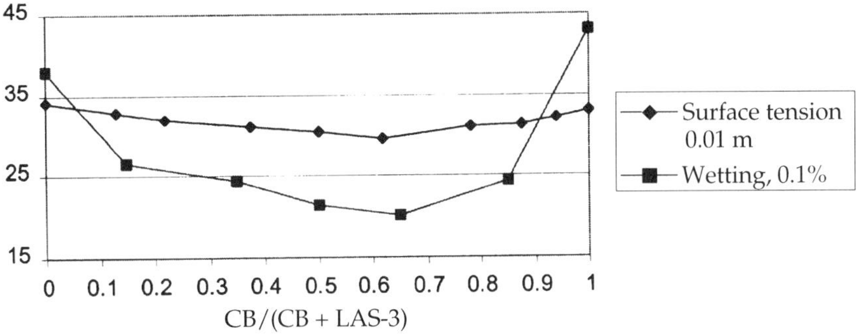

Figure 5.5. Synergism in surface tension reduction and wetting times for cocoamidopropylbetaine (CB)/lauryl ethoxysulfate (3OE) (LAS-3) mixtures at NaCl = 0.1 M, pH = 6, T = 25°C.

Recently, the understanding of synergism and the interaction of surfactants in mixtures has been made use of extensively in mitigating the irritancy of formulations containing highly irritating anionic and cationic surfactants. Alcohol sulfates (and their ether sulfates) and quaternary ammonium salts have long been used in the formulation of personal care and disinfecting formulations, respectively. *In vitro* Eyetex (1, 2) (Ropak Laboratories, Irvine, California) irritancy measurements for mixtures of lauryl alcohol (ether) sulfates/cocoamphoacetate or diacetate (3) are shown in Figures 5.7 and 5.8 (4, 5). As the molar ratio of amphoacetate is increased in each mixture, the irritancy—as measured by the Eyetex

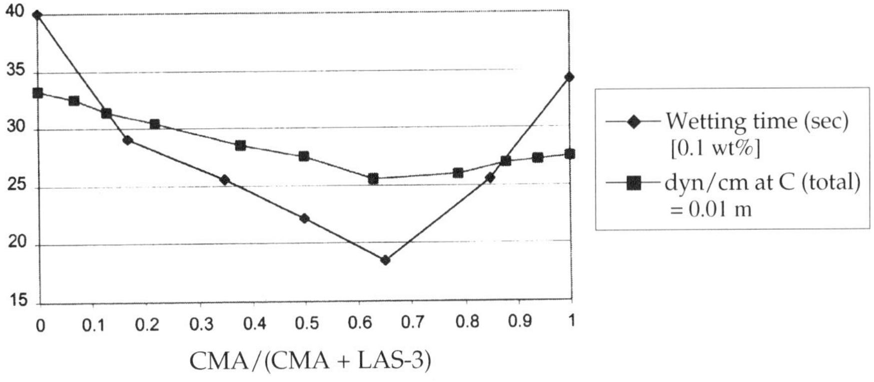

Figure 5.6. Synergism in surface tension reduction and wetting times for $C_{12}H_{25}C(O)NHCH_2CH_2N^+H(CH_2CH_2OH)CH_2COO^-$/lauryl ethoxysulfate (3OE) mixtures at NaCl = 0.1 M, pH = 6, T = 25°C.

TABLE 5.3
Synergism in Foam Enhancement for Some Zwitterionic and Anionic Surfactant Mixtures

Surfactant mixture	Foam height (mm)
$C_{12}SO_4^-Na^+$	199
$C_{12}(OE)_3SO_4^-Na^+$	189
$C_{12}Amp^+N(CH_3)_2CH_2COO^-$	186
$C_{12}N^+(CH_3)_2O^-$	185
$C_{12}Am^+N(E_t)(CH_2COO)_2$	187
$C_{12}Am^+N(E_t)(CH_2COO)_2/C_{12}(OE)_3SO_4^-Na^+$	198
$C_{12}Amp^+N(CH_3)_2CH_2COO^-/C_{12}C_6H_4SO_3^-Na^+$	209
$C_{12}N^+H(CH_3)_2O^-/C_{12}(OE)_3SO_3^-Na^+$	206
$C_{12}N^+H(CH_3)_2O^-/C_{12}SO_4^-Na^+$	204

Draize Equivalents (EDE) value for the anionic surfactant (and correlated as well to *in vivo* irritancy tests)—is dramatically reduced, with the maximum reduction in irritancy reached at the molar ratio corresponding to the maximum reduction in cmc or the lowest anionic monomer concentration (Table 5.4 and Fig. 5.9). The surfactant monomer concentration is thought to be responsible for ocular and skin irritation in human beings.

Gemini Surfactants

Gemini surfactants (6, 7) are compounds that contain two hydrophilic and two hydrophobic groups in the molecule, in contrast to conventional surfactants that contain only one hydrophilic and one hydrophobic group.

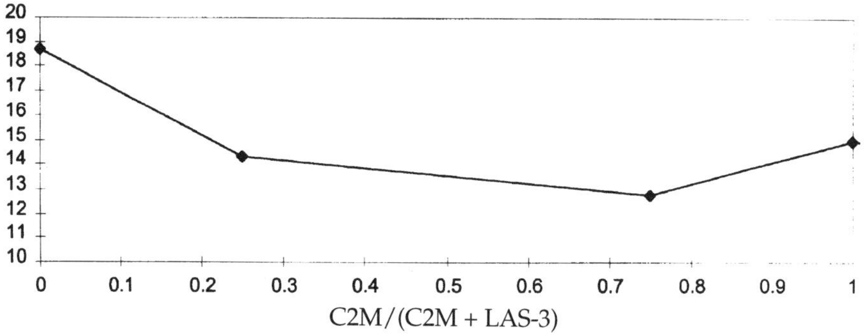

Figure 5.7. *In vitro* irritancy study on cocoamphoglycinate (C2M)[a]/lauryl ethoxysulfate (LAS-3) (3OE) at NaCl = 0.1 M, pH = 6, T = 25°C. [a]$C_{11}H_{23}C(O)NH(CH_2)_2N^+H(CH_2CH_2OH)CH_2COO^-$.

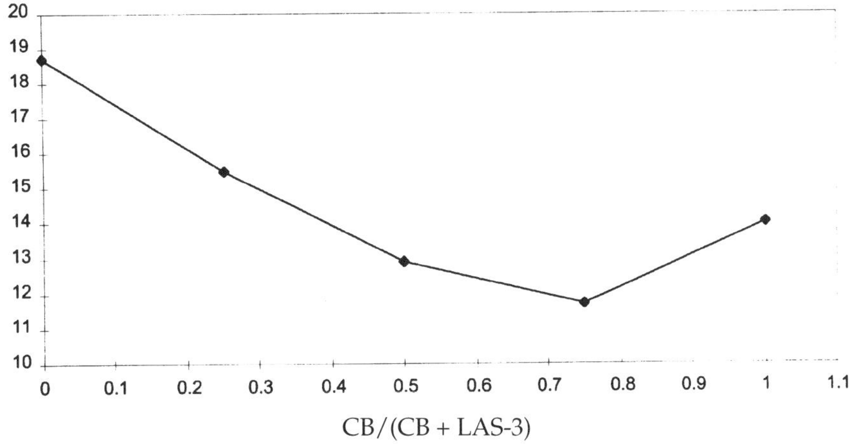

Figure 5.8. *In vitro* irritancy study on cocoamidopropylbetaine (CB)[a]/lauryl ether sulfate (LAS-3) (3OE) at NaCl = 0.1 M, pH = 6, T = 25°C.
[a]$C_{11}H_{23}C(O)N(CH_2)_3\ N^+(CH_3)_2{}^-CH_2COO^-$.

The two hydrophilic groups in the gemini surfactant are separated by a "spacer" or "linkage" containing two or more carbon atoms.

Gemini surfactants, sometimes called dimeric surfactants (3), are much more surface active, i.e., have much smaller cmc and C_{20} values, in

TABLE 5.4
Changes in Monomer Concentration of Sodium Lauryl Ethoxysulfate [$C_{12}(EO)_1S$] with Mole Fraction of Cocoamphoacetate (CMA)[a] in CMA/$C_{12}(EO)_1S$ (ESY) Mixtures (pH = 6.0, T = 25°C, in H_2O, C = 1.5 H 10^{-3} M)

Surfactant system	$\alpha_1{}^b$	cmc (M)	Monomer concentration of ESY (M)
$C_{12}(EO)_1S$	0.0	1.5×10^{-3}	1.5×10^{-3}
CMA/$C_{12}(EO)_1S$	0.1	2.9×10^{-4}	2.6×10^{-4}
CMA/$C_{12}(EO)_1S$	0.2	2.0×10^{-4}	1.6×10^{-4}
CMA/$C_{12}(EO)_1S$	0.3	1.5×10^{-4}	1.1×10^{-4}
CMA/$C_{12}(EO)_1S$	0.4	1.2×10^{-4}	7.2×10^{-5}
CMA/$C_{12}(EO)_1S$	0.5	1.0×10^{-4}	5.0×10^{-5}
CMA/$C_{12}(EO)_1S$	0.6	8.5×10^{-4}	3.4×10^{-4}
CMA/$C_{12}(EO)_1S$	0.7	7.8×10^{-4}	2.3×10^{-5}
CMA/$C_{12}(EO)_1S$	0.8	2.1×10^{-4}	4.2×10^{-5}
CMA/$C_{12}(EO)_1S$	0.9	26.1×10^{-4}	6.1×10^{-5}
CMA	1.0	9.8×10^{-4}	0

[a]CMA, $C_{12}H_{25}CONHCH_2CH_2N^+H(CH_2CH_2OH)CH_2OO^-$.
[b]Mole fraction of CMA in mixed solution.

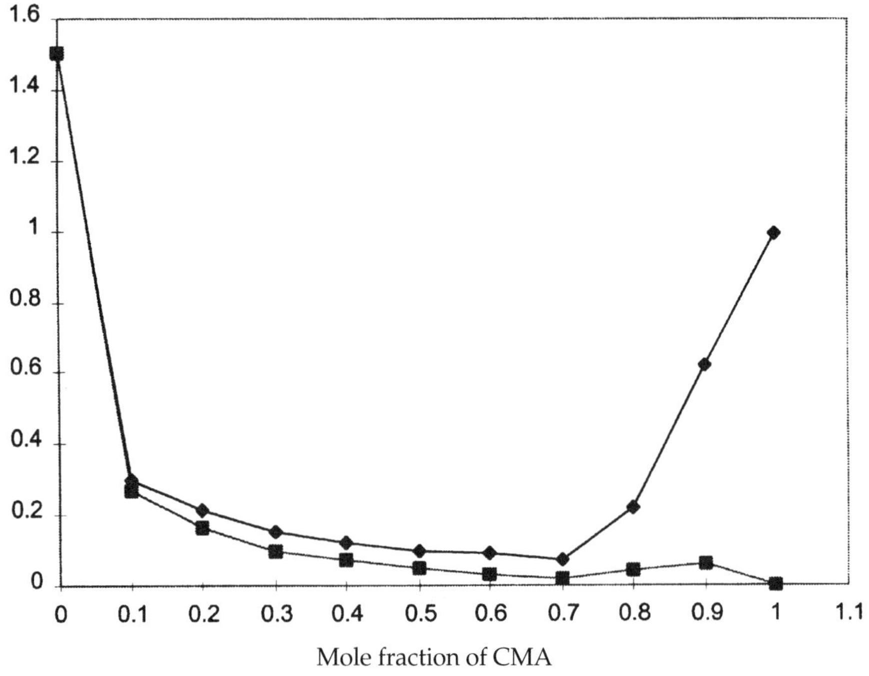

Figure 5.9. Dependence of mixed cmc and monomer concentration of lauryl ethoxysulfate (LAS-3) (3OE) on the mole fraction of cocoamphoacetate (CMA)/LAS-3 mixtures (pH = 6, T = 25°C, in H_2O).

aqueous media than conventional surfactants containing similar single hydrophilic and hydrophobic groups. This is because surface activity depends on the distortion of the solvent (water) structure by the hydrophobic group(s) of the surfactant molecule. The two hydrophobic groups of the gemini distort the water structure more than does the single hydrophobic group of the conventional surfactant. The second hydrophilic group of the gemini serves to keep it water soluble in spite of the presence of the two hydrophobic groups in the molecule.

Having cmc and C_{20} values up to 1,000 times smaller than conventional surfactants with single similar hydrophilic and hydrophobic groups, gemini surfactants are much more efficient both at forming micelles and at adsorbing at interfaces. They also show closer packing of their hydrophobic groups at the water/air interface, with consequent greater effectiveness of surface tension reduction and greater cohesiveness of their interfacial films than comparable conventional surfactants. Ionic geminis, because of their double charge, show stronger interaction with oppositely charged surfactants, with consequent greater probability of synergy, and are also more soluble in water than comporable conventional ones. Some geminis show unique rheological properties in aqueous solution. Thus, some catonics form long, thread- or worm-like micelles at low concentrations in aqueous solution, showing shear thickening and then viscoelasticity at slightly higher concentrations (8). There has been intense interest in geminis since their unusual surface activity was first pointed out several years ago (2), with a flood of papers in the scientific literature and patents, which have recently been reviewed (9).

The effect of structural changes in the gemini molecule on fundamental properties has been covered in numerous studies. These indicate that surface activity in aqueous media is greatest (i.e., has the smallest cmc and C_{20} values) when the linkage between the two hydrophilic head groups is short, capable of hydrogen bonding to water, and flexible. Surface activity increases with an increase in the length of the alkyl chains in the hydrophobic groups in the gemini molecule in a manner similar to that observed in conventional surfactants but, in contrast to that observed in the latter, starts to deviate from this pattern when the carbon atoms in each alkyl chain of the hydrophobic groups exceed a certain number (often >14). The deviation becomes greater and greater with further increase in the number of carbons and results in a *decrease* in surface activity with further increase in the number of carbon actions in the alkyl chain (10, 11, 12). This behavior has been attributed to the formation of small, non-surface-active premicellar aggregates (dimers, trimers), whose presence has been confirmed by a study of their aggregation numbers (13).

Since gemini surfactants are a relatively new type of surfactant, few types are currently commercially available, although one type, the acetylenic glycols

of structure 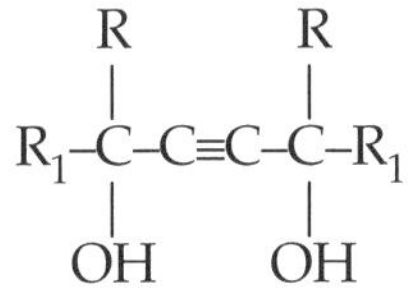 (14), have been used as nonfoaming wetting agents for decades, and experimental quantities of disodium dialkyldiphenyl ether disulfonates, of the following structure

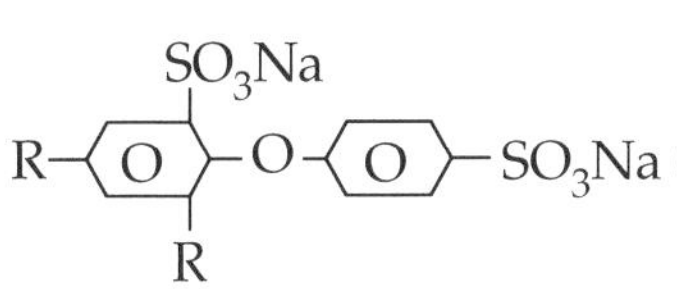 (15), are available:

The unique properties of gemini surfactants mentioned above, indicating great potential for industrial applications have resulted in numerous scientific investigations and patents involving their use, however. Their greater foaming ability, compared to conventional surfactants with similar single hydrophilic and hydrophobic groups, has been observed by several investigators (16–20). This includes cationic geminis (20), which show high foaming ability in contrast to conventional cationics that show very little foaming. Besides the commercially available acetylenic glycols mentioned above, other geminis have been observed to be excellent wetting agents. These include the dicarboxylates of structure $[C_{10}H_{21}OCH_2CH_2(OCH_2\ COO–Na^+)\ CH_2]_2$ Y, where Y is –O– or $–OCH_2CH_2O–$, as textile wetting agents (15) and the nonionic diamides of structure $\{C_4H_9CH(C_2H_5)CH_2N[C_2H_4O)_4H]C(O)\}_2$ Y, where Y is $–CH_2CH_2–$ or –CH=CH–, as hydrophobic soil wetting and rewetting agents (21).

Gemini surfactants have also been found to solubilize water-insoluble substances, including hydrocarbons, water-insoluble surfactants, and water-insoluble pollutants more efficiently and more effectively than comparable conventional surfactants. Thus, the cationic gemini of structure $[C_{12}H_{25}N^+(CH_3)_2\ CH_2CH_2]_2\ Y \cdot 2Br^-$, where Y is $–CH_2–$, solubilizes *n*-hexane and toluene better than the comparable conventional surfactant (22), anionic geminis of type disodium didecyl diphenyl ether disulfonate solubilize water-insoluble nonionic surfactants more efficiently and effectively than the comparable monosodium monodecyl diphenyl ether monosulfonate (23), and the cationic geminis of structure $[RN+(CH_3)_2\ CH_2CH_2]_2 \cdot 2Br^-$, where R is $C_{12}H_{25}$ or $C_{14}H_{29}$, adsorb more efficiently on clay and remove the pollutants 2-naphthol and 4-chlorophenol more efficiently and effectively from aqueous media than the corresponding alkyl trimethylammonium bromides, when adsorbed on the clay (24).

Surfactants of the gemini type have also been found to have superior antimicrobial properties, as well as to show less skin irritancy. The dicationic of structure $[C_{12}H_{25}N^+(CH_3)_2\ CH_2CONH_2]_2\ Y \cdot 2CH$, where Y is $–(CH_2)_4^-$ or $–(CH_2)_2\ S\ S(CH_2)_2^-$, shows greater antimicrobial activity than hexadecyltrimethylammonium bromide against both gram-positive and gram-negative organisms and *Candida albicans* (25). The low skin irritancy of geminis, including dicationics, is probably is due to their low cmc values, because skin irritancy is generally due to the monomeric surfactant concentration.

Other Methods of Enhancing Performance

Performance properties of surfactants can sometimes be enhanced by less obvious methods. Water-insoluble surfactants may produce very low surface tensions in aqueous solution and still show very poor wetting speeds for textile surfaces. This is probably because their very low solubility in water results in very slow rates of surface tension reduction. The addition of a surfactant that increases the solubility of the water-insoluble surfactant in the aqueous phase can result in much faster wetting. This is possible even when the water-soluble surfactant, by itself, shows very poor wetting speed. A related example is the increase in foaming that occurs when a solubilizing surfactant is added to a second surfactant that has only borderline solubility in water.

A mixture of a low HLB (hydrophilic–lipophilic balance) and a high HLB polyoxyethylenated (ethoxylated) nonionic surfactant is well known to produce a more stable oil-in-water emulsion than a single, similar polyoxyethylenated nonionic surfactant with the same weighted average HLB as the mixture. Similarly, emulsification with a polyoxyethylenated nonionic surfactant is improved when the surfactant has a wide, rather than a narrow, distribution of polyoxyethylene chain lengths. A low HLB surfactant of this type has a small hydrophilic head group consisting of a polyoxyethylene chain with a small number of oxyethylene units; a high HLB surfactant of this type has a large hydrophilic head group, consisting of a polyoxyethylene chain with a relatively large number of oxyethylene units. The long polyoxyethylene chain of the high HLB molecule adsorbed at the O/W interface extends into the water phase and produces a steric barrier to the coalescence of the dispersed oil particles in the O/W emulsion. However, the large head group of the high HLB molecule prevents these molecules adsorbed at the interface from approaching each other closely. The interfacial film produced from these molecules alone would have rather weak hydrophobic chain-chain lateral attraction because of this loose packing and would consequently be unstable, producing an unstable emulsion. The low HLB molecules, with their small hydrophilic heads, are intercalated at the O/W interface in the spaces between the high HLB molecules, increasing the chain-chain cohesiveness of the interfacial film, and consequently the stability of the emulsion.

Mixtures of silicone-based (trisiloxane) surfactants and certain short-chain hydrocarbon surfactants are extensively used in industry (as coatings and adjuvants for agrochemicals) to enhance the superspreading properties of aqueous solutions on hydrophobic surfaces. Spreading factors (i.e., spreading area of 0.008 g of 0.17 (by wt) surfactant solution relative to the area spread by 0.008 g of deionized water on Parafilm after 1 min) for a silicone (trisiloxane) surfactant, some conventional surfactants, and their mixtures that are used industrially are shown in Table 5.5.

Interactions between cationic surfactants and zwitterionic surfactants are made use of in disinfectant formulations. In general, the traditional disinfectant formulations, a majority of which are based on mixtures of quaternary ammonium salts and nonionic surfactants, are highly toxic. At the same time, they provide poor detergency (due to the poor detergency of the quaternaries), which has been a major drawback with disinfectant formulations. In order to overcome this deficiency, formulators are now replacing nonionic surfactants with amphoterics that are compatible and that interact with the biocidal quaternary ammonium salt, thereby enhancing the detergency and reducing the toxicity while at the same time maintaining the antimicrobial properties of the quaternaries (Table 5.6). Anionics on the other hand, although they interact much more strongly with cationic surfactants, are usually

TABLE 5.5
Superspreading Performance of Silicone and Some Hydrocarbon-Based Surfactants and Their Mixtures on Parafilm

Surfactant system (0.1% by wt)	Spreading factor
Silwet L-77[a]	15.0
Sodium 2-ethylhexyl sulfate	3.6
Sodium octyl sulfate	6.0
Sodium di-isopropyl saphthalene sulfonate	3.0
Sodium C_8-amphocarboxylate	2.5
Dimethyl octynediol	6.5
n-Octyl-2-pyrrolidone	7.5
Silwet L-77/ sodium 2-ethylhexyl sulfate (9:1)	17.5
Silwet L-77/ sodium octyl sulfate (9:1)	25.0
Silwet L-77/ Sodium di-isopropyl naphthalene sulfonate (9:1)	30.0
Silwet L-77/sodium C_8–C_{10} amphocarboxylate[b] (9:1)	25.0
Silwet L-77/ dimethyl octynediol (3:1)	30.0
Silwet L-77/ *n*-octyl-2-pyrrolidone	30.0

[a]Abbreviations: $(CH_3)_3$ Si–O–Si$(CH_3)[OC_3H_6(OC_2H_4)_{7.5}OCH_3]OSi(CH_3)_3$ (Crompton Corp.).
[b]Amphocarboxylate, $C(O)NHCH_2CH_2{}^+NH(CH_2CH_2COO^-)CH_2$ CH_2COOH.

TABLE 5.6
Irritancy (Score) and Hard Surface Cleaning Performance (HSC) of Cationic/Amphoteric Mixtures, pH = 6

Formula	Score[a]	HSC% soil removal	β
$C_{16}{}^{+}N(CH_3)_3Cl^-$	52.4	35	—
$C_{16}{}^{+}N(CH_3)_3Cl^-/C_{12}N^+HCH_2CH_2COO^-$ (1.6:1)	18.2	65	−3.2
$C_{16}{}^{+}N(CH_3)_3Cl^-/C_{12}Am^+N(Et)(CH_2CH_2COO^-)_2$ (1.6:1)	17.5	62	−4.1

[a]Eye irritancy, rabbit. Maximum possible score = 100. Use concentration, 0.1 mL of 10% solution.

incompatible and also destroy the antimicrobial properties of the quaternaries, when used with them.

References

1. Kary, J.H., and J.C. Callandra, *J. Soc. Cos. Chem. 13:*281–289 (1962).
2. Gordon, V.C., and H.C. Bergman, in Alt. Methods in Tox. edited by A.M. Goldberg, Mary Ann Liebert, New York, 1983, pp. 293–302.
3. Dahanayake, M., T. Gao, and R. Vukov, "Synergistic Interaction and Performance Properties Of Cocoamphoglycinates and Aklyethersulfates," *4th CESIO International World Surfactants Congress and Exhibition*, Barcelona, Spain, Section D, 1996, pp. 299–318.
4. Zana, R., and Y. Talman, *Nature 362:*228 (1993).
5. Danino, D., Y. Talman, and R. Zana, *Langmuir 11:*1448 (1995).
6. Menger, F., and C.A. Littau, *J. Am. Chem. Soc. 113:*1451 (1991).
7. Rosen, M.J., *CHEMTECH 23:*30 (1993).
8. Rosen, M.J., and D.J. Tracy, *J. Surfactants Detergents 1:*547 (1998).
9. Menger, F.M., and J.S. Keiper, *Angew. Chem. Int. Ed. 39:*1906 (2000).
10. Menger, F., and C.A. Littau, *J. Am. Chem. Soc. 115:*10083 (1993).
11. Sung, L.D., and M.J. Rosen, *Langmuir 12:*1149 (1996).
12. Rosen, M.J., and L. Liu, *J. Am. Oil Chem. Soc. 73:*885 (1996).
13. Mathias, J.H., M.J. Rosen, and L. Davenport, *Langmuir 15:*7340 (1999), and unpublished data.
14. Air Products and Chemicals, Allentown, Pensylvannia.
15. Dow Chemical Co., Midland, Michigan.
16. Zhu, Y.-P., A. Masuyama, T. Nagata, and M. Okahara, *J. Jpn. Oil. Chem. Soc. (Yukagaku) 40:*473 (1991).
17. Ono, D., T. Tanaka, A. Masuyama, Y. Nakatsuji, and M. Okahara, *J. Jpn. Oil. Chem. Soc.* (Yukagaku) *42:*10 (1993).
18. Zhu, Y.-P., A. Masuyama, Y. Kobata, Y. Nakatsuji, M. Okahara, and M.J. Rosen, *J. Colloid Interface Sci. 158:*40 (1993).
19. Zhu, Y.-P., A. Masuyama, and M. Okahara, *J. Am. Oil Chem. Soc. 68:*268 (1991).

20. Kim, T.-S., T. Kida, Y. Nakatsuji, T. Hirao, and I. Ikeda, *J. Am. Oil Chem. Soc. 73:*907 (1996).
21. Micich, T.J., and W.M. Linfield, *J. Am. Oil Chem. Soc. 65:*820 (1988).
22. Dam, T., J.B.F.N. Engberts, J. Karthauser, S. Karaborni, and N.M. VanOs, *Colloids Surfs. A 118:*41 (1992).
23. Rosen, M.J., Z.H. Zhu, and X.Y. Hua, *J. Am. Oil Chem. Soc. 69:*30 (1992).
24. Li, F., and M.J. Rosen, *J. Colloid Interface Sci. 224:*265 (2000).
25. Diz, M.A., A. Pinayo, P. Erra, and M.R. Infante, *J. Chem. Soc., Perkin Trans. 2:*1871 (1994).

CHAPTER 6

Surfactant Applications 1

Agrochemicals

Using Adjuvants to Enhance Wetting or Spreading on the Substrate

Adjuvants are chemicals that increase the effects of the biological actives in agrochemical preparations (1). Surfactants can play a major role in enhancing the effectiveness of the biologically active ingredients (2–4). Although the role of the surfactant adjuvants in enhancing their effectiveness is far from being understood in all cases, a clear relationship exists between enhanced wetting of or spreading over the substrate and herbicidal efficacy. In some cases, this is accomplished by ion pairing or complexing of biologically active material, usually with improved wetting or spreading. This enhanced wetting or spreading can significantly reduce the amount of active ingredient required to produce a given biological effect.

Since in practice sufficient time exists for spreading or wetting to reach equilibrium, the processes are determined by the equilibrium spreading coefficient (Eq. 2.8); consequently, the surfactant properties desired are reduction of the surface tension of the solution and reduction of the substrate/solution interfacial tension. Since the substrates are mostly hydrophobic in nature, surfactants that will adsorb onto the substrate in such fashion as to lower the substrate/solution interfacial tension (Chapter 2, Changes in the Properties of Solid/Liquid and Liquid/Liquid Interfaces) in addition to reducing the surface tension of the solution (Chapter 2, Changes in the Properties of the Surface of a Solution) are desired. Since substrate/liquid interfacial tension is not readily measured, surfactants are used that reduce the surface tension of the solution to a very low value (in almost all cases, aqueous) and that have a long hydrophobic group capable of adsorbing well on hydrophobic substrates. (Table 6.1 lists surface tension values for some typical surfactants.)

Most of the surfactants used for this purpose are nonionic and have branched hydrophobes (5,6). The typical HLB value for these compounds ranges from 9 to 14, and the branched hydrophobes yield lower surface tensions than linear-chain analogs. Generally, the surfactants are used below or close to their CMC to prevent solubilization of active material, which would decrease their biological activity. Anionics foam excessively and are seldom used in these applications, even when they provide good equilibrium wetting properties. Also, since most of the adjuvants are sprayed, spray drift is a major problem; those surfactants that provide good dynamic surface tension reductions would produce very small particles that would enhance drift. Therefore, good dynamic wetting agents, such as short-chain nonionic surfactants and anionic

TABLE 6.1
Minimum Aqueous Solution Surface Tension Values for Some Typical Surfactants

Surfactant	Minimum equilibrium surface tension (dyn cm^{-1})
Nonylphenol ethoxylate (9 EO)	29.1
Lauryl alcohol ethoxylate (9 EO)	32.2
Tall oil fatty acid ethoxylate (10 EO)	33.5
Tridecyl alcohol ethoxylate (8–9 EO)	27.5
Sodium di (2-ethylhexyl) sulfosuccinate	25.5
Trisiloxane ethoxylate (8 EO),CH_3—terminated	21.0

surfactants, are generally avoided. Trisiloxane-based surfactants are very useful for this purpose, since they can reduce the equilibrium surface tension to values (20 dyn/cm) much lower than the minimum value (25 dyn/cm) attainable with hydrocarbon-chain surfactants, and since they are not very good dynamic wetting agents. They spread much better than hydrocarbon-chain surfactants on hydrophobic surfaces such as plant leaves (8–12). Also, very often mixtures of trisiloxane with branched and short hydrocarbon anionic surfactants are used due to their synergetic interactions in enhancing superspreading properties (13–15). Superspreading performance for some of these mixtures of surfactants are shown in Table 5.5.

Wettable Powders

Wettable powders are prepared by blending surfactants and finely powdered fillers with either liquid or solid active ingredients. These wettable powders are diluted with water, and the well-dispersed suspension is sprayed. In most cases, the active ingredient constitutes 10 to 60% of the formulation. Since these actives are generally difficult to wet with water by itself, they need to be formulated with good wetting agents. The wetting agents should have the ability to adsorb onto the solid well and provide instantaneous wetting and dispersibility to the powder when it is added to the spray tank. The suspension stability required is fairly short: 2 min at the longest and 1 min on average.

Dispersability and wettability are the most important interfacial properties that are required in the formulation of wettable powders. The wetting agents must provide instantaneous wetting once the powders are diluted with water. In addition to wetting, the surfactants must provide good temporary dispersibility for the powders to maintain homogeneity of the suspension in hard and cold water. In most cases, powdered surfactants are preferred over liquid surfactants because of their greater ease of incorporation.

Surfactants that are commonly used to provide dispersibility are the sodium alkyl naphthalene sulfonates, aryl-formaldehyde condensates, lignosulfonates, and ethoxylated tristyrylphenol sulfates (16–17). In general, these are surfactants that have short, aromatic nuclei-containing hydrophobes, and multiple ionic groups. The aromatic nuclei in the hydrophobic portion of the molecule are polarizable and adsorb well, with the proper orientation (of the hydrophobic portion toward the substrate and the hydrophilic groups toward the water), onto the actives, which also have polarizable groups in the molecule. This orientation of the adsorbed surfactant decreases the solid/water interfacial tension. The multiple ionic groups in the surfactant molecule promote dispersibility of the actives in water, even cold and hard water. Most of the wetting properties of the surfactant formulation, however, are provided by the addition of C12 alkyl sulfates, dioctylsulfosuccinates, and alkylbenzenesulfonates that are used at very low concentrations compared to the shorter chain naphthalene sulfates, lignosulfonates, and other aromatic sulfonates and sulfates that are used for their dispersion properties. Carboxylates and phosphates are rarely used, as these are often liquids and difficult to handle.

Caking under humid, high-temperature conditions is a problem; therefore, the surfactant used should not be hygroscopic and should provide anticaking properties. Thus, ethoxylated nonionic surfactants that are hygroscopic are rarely used in the formulation of wettable powders.

Compared to suspension concentrates (discussed below), only small amounts of surfactants are present in wettable powders. The function of the surfactants is, most importantly, to keep the solids suspended in the final product just before spray application. Table 6.2 shows the major functions of surfactants commonly used in wettable powders.

Suspension Concentrates

The availability of suspension concentrates (SC), commonly called "flowables," has increased rapidly in recent times. They are prepared by dispersing a finely powdered, water-insoluble, solid active ingredient in water; therefore, they are similar to dispersions of wettable powders. However, the need for physico-chemical stability of the dispersion in water is very much greater than for wettable powders. They are highly concentrated suspensions in water (30 to 60% actives), and must be free from precipitation, hydrolysis, and flocculation over a longer period of time than dispersions of wettable powders. Furthermore,

TABLE 6.2
Surfactants Used in Wettable Powders

Surfactants	Major function
Dodecylbenzene sulfonates	Wetting
Dioctylsulfosuccinates	Wetting
Sodium alkylnaphthalene sulfonates	Wetting
Naphthalenesulfonate–formaldehyde condensates	Dispersing
Ethoxylated tristyrylphenol sulfates	Dispersing
Sodium lignosulfonates	Dispersing

when diluted with water prior to spraying, the flowables must disperse rapidly and uniformly in water.

Suspension concentrates have five advantages over wettable powders:

1. They are easier to use than wettable powders, because they are already dispersed in water.
2. They are easier to disperse prior to application.
3. They produce no dust in use.
4. They have a lower packing volume.
5. They are easily dilutable and produce more stable suspension forms on dilution with water.

The actives formulated as SCs should have very low water solubility, as these products must have good shelf stability and the actives must not precipitate *via* recrystallization. This property requires the use of very good dispersing agents, as well as wetting agents, in their formulations. Whereas surfactants with good dispersing properties (relative to wetting properties) are generally used for the formulations of wettable powders, in the selection of surfactants for suspension concentrates, surfactants having both good (equilibrium) wetting and dispersing properties are used, and at concentrations greater than for wettable powders.

In the majority of applications, the wetting agents selected are anionics with a high charge density per molecule. Anionic sulfonates whose hydrophobic groups adsorb onto the actives to produce (presumably) very low solid/liquid interfacial tensions and whose ionic head groups produce good electrostatic stability are generally used. Low solid/liquid interfacial tension resulting from the adsorption of these surfactants onto the actives will provide good wettability to the actives, since they are generally intrinsically hard to wet. The electrical charges on them as a result of the adsorbed anionic surfactant molecules will give them good dispersibility in water. Dialkyl sulfosuccinates and alkyl diphenyl ether disulfonates are the most commonly used. The sulfonate groups in alkyl diphenyl ether disulfonates provide excellent dispersing power and prevent crystal growth in the actives. In general, molecules having aromatic nuclei or olefinic bonds in the hydrophobic group are preferred, as they provide good substantivity to the actives (see Table 6.3). In recent times tristyrylphenol sulfates and phosphates, due to their excellent dispersing and wetting properties are increasingly being used (17, 18). Nonionic surfactants are used only when incompatibility pre-

TABLE 6.3
Surfactants Used in Suspension Concentrates

Surfactant	Function
Sodium dodecylbenzene sulfonates	Wetting
Dibutyl and di-isopropyl naphthalene sulfonates	Wetting
Dioctyl or dinonylphenolsulfosuccinates	Wetting
N-methyl oleyl taurates	Wetting and dispersing
Naphthalene sulfonate–formaldehyde condensates	Dispersing
Lignosulfonates	Dispersing
Dodecyldiphenylether disulfonates	Dispersing
Ethoxylated (6–12 EO) nonylphenol phosphate esters	Wetting and dispersing
Ethoxylated (14–16 EO) tristyrlphenol phosphate sodium salt	Wetting and dispersing

vents the use of anionic surfactants, since the former do not produce electrostatic barriers to coalescence. For both anionics and nonionic surfactants in general, branched-chain alkyl groups are preferred over linear ones, as these provide a greater lowering of interfacial tension at the solid/liquid and air/liquid interfaces. Also, branched surfactants provide good freeze–thaw stability and much lower viscosity for the suspensions than linear-chain hydrophobes.

Emulsion Polymerization

Surfactants are indispensable ingredients in the emulsion polymerization process, where they serve as both monomer emulsifiers and latex stabilizers.

The emulsion polymerization process involves the polymerization of liquid monomers dispersed in an aqueous surfactant phase. This dispersed monomer, in an exothermic reaction, yields polymer particles ranging in size from 0.05 to 5.0 mm. The aqueous surfactant phase is required both for emulsification and as a heat transfer medium. The amount of water used determines the total solids of the emulsion. Commercially, the total solids of an emulsion are maintained at as high a level as possible, consistent with the requirements of heat removal, viscosity, and stability of the final product.

The mechanism of emulsion polymerization and the role played by surfactants in emulsion polymerization, although the subject of extensive research, still remain controversial. The micellar theory of monomer solubilization by surfactant micelles, although widely believed, has little evidence to support it. This theory is based on the assumption that the micelles generated by the surfactants solubilize the monomers and provide a site for polymerization. According to this theory, the size of the particles formed is dependent on the efficiency of micelle formation; the lower the CMC of the surfactant, the smaller the particle size of the emulsion. This theory cannot be justified, however, as the weight fraction of the monomer typically used commercially in emulsion polymerization far exceeds the limit of solubilization possible by the total number of micelles formed, even for a very efficient surfactant. In addition, it is very well known that anionic surfactants, which give smaller and more uniform particle sizes in emulsion polymerization than nonionic surfactants, are not better solubilizers than the latter, which are generally the best solubilizers for monomers. Since relatively small levels of surfactants are used in the emulsion polymerization reaction (3 to 6%, based on monomer content), it appears that emulsification rather than micelle formation plays the major role in the emulsion polymerization process. Consequently, reduction of the monomer/aqueous solution and polymer/aqueous solution interfacial tensions, which facilitate emulsification, are some of the desired properties of the surfactants, and since in emulsion polymerization sufficient time exists for the emulsion to reach equilibrium conditions, the desired surfactant properties are low equilibrium tensions at those interfaces. Surfactants that can adsorb well at the hydrophobic monomer/aqueous interface and the growing polymer/aqueous interface, with minimum solubility in either phase so as to efficiently lower the interfacial tensions there, are desired.

Another very important interfacial property of the surfactant is to provide stability to the dispersion of the growing polymer particles. Stabilization of the dispersion of the polymer particles is critical to generating smaller and uniform particle sizes, to preventing coagulation, and also for providing both mechanical and chemical stability to the dispersion.

Unlike the situation in most other industrial applications, the surfactants used in emulsion polymerization remain part of the polymer particles and significantly influence the physical and interfacial properties of the polymer. The presence of an adsorbed layer of surfactant molecules on the surface of the polymer can affect such properties of the polymer as its sensitivity to water and other environmental components, its wettability by solvents, and its adhesion to other substrates. It can also affect such film properties as gloss, clarity, and mechanical and shelf stability. Therefore, in the selection of surfactants for emulsion polymerization, considerable attention has to be given to the effect of the surfactant on the properties of the end product.

Although both anionic and nonionic surfactants are used, the most commonly used primary surfactants for emulsion polymerization are the anionic surfactants. The ionic groups in the surfactant molecule promote stability for the monomer/water emulsions and more importantly, they control the particle size distribution by stabilizing of the dispersion of the particles. Moreover, polyoxyethylenated nonionics, compared to anionics, produce greater water sensitivity in the final polymers, which in certain situations is detrimental to their application.

In order to provide maximum reduction in monomer/aqueous solution (and air/aqueous) interfacial tensions, the surfactants used have hydrophobes that are long and mostly branched, and that very often contain polarizable groups such as aryl groups. The polarizable groups adsorb well, with the proper orientation, onto the monomers and growing polymer particles, thereby decreasing monomer/aqueous solution and polymer/aqueous solution interfacial tensions. Branched hydrophobes, in addition to providing good interfacial tension reduction, provide good solubility for the surfactant in the presence of polycations as well as good freeze–thaw and shear stability to the dispersions.

Among the anionics, sodium alkyl (branched) benzene sulfonates, alkyl-diphenyl ether disulfonates, and sodium polyoxyethylene alkyl and polyoxyethylene alkyl sulfates are the most extensively used surfactants in emulsion polymerization. As is generally the case in emulsification (Chapter 4, Dispersion of Solids in Liquid Media), the hydrophile–lipophile balance (HLB) of the surfactants that are best for the emulsification of the monomer roughly reflects the HLB of the latter. In the case of the relatively more hydrophobic monomers (such as the butylacrylates, methacrylates, styrene, and styrene/butyl acrylate mixtures), sodium lauryl sulfate, sodium dodecylbenzene sulfonates, dodecyldiphenyl ether disulfonates, and alkyl phenol ether sulfates with 4 to 15 mol of EO are used. Diphenyl ether disulfonates, due to the high charge density in the molecule, provide excellent electrostatic stability with the more hydrophobic monomers and therefore are commonly used in these applications. In the case of the more hydrophilic monomers, such as vinyl acetate and vinyl acrylates, more hydrophilic anionics, such as the alkyl aryl ether sulfates having polyoxyethylene chains ranging from 10 to 40 mol, are commonly used. Table 6.4 lists surfactants used for various monomers.

TABLE 6.4
Surfactants Used for Various Monomers

Monomer	Surfactant used
Styrene, butadiene	Dodecylbenzene sulfonates, dodecyldiphenyl ether disulfonates
Vinyl chloride	Partially hydrogenated fatty acid soaps, sodium lauryl sulfate
Styrene/butadiene	Fatty acid soaps, dodecylbenzene sulfonates, dodecyldiphenyl ether disulfonates, polyoxyethylenated (9–14 EO) octylphenols
Methyl or butyl esters of acrylic/methacrylic acid	Sodium lauryl sulfate octyl, nonylphenol ether (4–10 EO) sulfates
Methacrylic acid/acrylic acid	Sodium C_{12}–C_{14} ether (4–8 EO) sulfates, dodecylbenzene sulfonates
Styrene/butyl acrylate	Nonyl/octyl phenol polyoxyethylene (9–15 EO) sulfate sodium polyoxyethylene (4–10 EO) lauryl ether sulfates, sodium or ammonium C_{12}–C_{14} ether (2–10 EO) sulfates, polyoxyethylene (30–50 EO) octyl/nonylphenols
Vinyl, vinyl acetate, vinyl acetate/butyl acrylate	Sodium polyoxyethylene (30–50 EO) nonylphenol ether sulfates, sodium polyoxyethylene (30–50 EO) lauryl ether sulfate, polyoxyethylene (30–50 EO) octyl/nonylphenols

In addition to the anionic surfactants, nonionic surfactants are also widely used in emulsion polymerization, but as secondary emulsifiers. They are rarely used alone, except in the case of very hydrophilic monomers where stabilizing the polymers is not a major problem. In selecting nonionic surfactants for emulsion polymerization, the HLB rule is very helpful. As the hydrophilicity of the monomer increases, a higher HLB nonionic surfactant is generally more effective. The majority of nonionic surfactants used have octyl- or nonylphenol-based hydrophobes. Moles of ethoxylation for these molecules range from 9 to 50. Among these, the lower mole ethoxylates (9 to 15 EO) are often used along with anionic surfactants to provide synergism in the reduction of interfacial tension essential for the efficient emulsification of the monomers. Ethoxylated octyl- and nonylphenols with 30 to 40 mol of EO are extensively used, along with their corresponding anionic sulfates, in the polymerization of both hydrophobic and hydrophilic monomers. Their main function in the emulsion polymerization process appears to be to provide steric stability via their long polyoxyethylene chains, i.e., by acting as protective colloids. Compared to the anionic surfactants used, these nonionics, due to their increased solubility at lower temperatures, ensure good freeze–thaw stabilization under storage conditions and in the presence of the multivalent cations encountered when they are formulated into paints and coatings. The required total surfactant concentration of an anionic emulsion stabilized with a nonionic surfactant will usually be lower than for a purely anionic stabilized emulsion.

Metal Cleaning

In cleaning metals, the cleaning system is chosen with consideration of the type of process metal contaminant to be removed and the washing process (immersion, spraying, etc.) to be used. Chemical cleaning is more widely used than mechanical cleaning in most industrial processes. Aqueous detergent systems often used in chemical methods include alkaline- or acid-based systems. In more critical areas, such as precision cleaning, or where the metal is subjected to a subsequent treatment process such a electroplating, annealing, or painting, the surface must be absolutely free of any residue after cleaning. Major cleaning processes can be classified broadly into two classes:

1. Immersion cleaning
2. Spray cleaning

The surfactants utilized in the formulation of cleaners for these two processes are very different.

Immersion Cleaning

Immersion cleaning is simple and is the most widely used. Here, the metal parts to be cleaned are soaked, with agitation, in a detergent solution for a considerable amount of time. In immersion cleaning, compared to spray cleaning, considerably less mechanical work is involved in the cleaning process. In the majority of cleaning processes, the parts to be cleaned are heavily soiled with oils, waxes, and greases, and the wash water in the cleaning process is reused for days or weeks. Therefore, performance demand from the surfactants used in metal cleaning is much more stringent compared to those used in laundry detergent and household cleaners.

Surfactant performance requirements for the formulation of high performance immersion cleaners are as follows:

- Good equilibrium wetting
- Effective reduction of surface and oil/water interfacial tensions
- Ability to form stable emulsions with or capacity to solubilize the hydrocarbon or fatty ester oily soils
- Prevention of the redeposition of soil
- Alkaline or acid stability and compatibility
- Corrosion inhibition

In immersion cleaning, considerable time exists for the wetting of surfaces to reach equilibrium; consequently, the most desired interfacial properties are the reduction of air/water and oil/water interfacial tensions to low equilibrium values, with a corresponding increase in (equilibrium) wetting (Chapter 2, Wetting and Dewetting). Therefore, surfactants with moderately long-chain hydrophobes, but with good solubility in the presence of high electrolyte or heavy metal ion concentrations, are used. Through their adsorption onto the soil/water interface, these surfactants produce good surface wetting of hard-to-wet oil-contaminated metal surfaces.

Because the contaminated wash water is reused and the oily soil buildup may reach as high as 20%, surfactants are used at much higher concentrations (2 to 4%) than the concentration used in laundry applications (0.1%). Since this is several orders of magnitude higher than the CMCs of the surfactants, effectiveness in reducing surface and oil/water interfacial tension reductions (i.e., the amount of reduction of the tensions) is considered more important than efficiency, in selecting surfactants.

For immersion cleaning, retention of the oil in the wash water through emulsification is extremely important in order to prevent its redeposition onto the metal surfaces. Consequently, the surfactants used must produce low oily soil/water interfacial tensions and provide good stability to the oily soil/water emulsions formed. At low soil load conditions, solubilization of the oils by the surfactants may also play a major role.

The majority of surfactants used in these applications are anionic or anionic/nonionic mixtures. Since the majority of metal cleaners are heavily built—either alkaline (pH >10) or acid (pH <4)—the surfactants used must have good solubility and stability in these systems. Nonionic surfactants in general, and anionics such as sulfonates, sulfates, and carboxylates, are either incompatible or unstable in these heavily built alkaline or acid systems. Low mole ethoxylated (4 to 10 EO) phosphate ester and amphoteric surfactants are commonly used, since they provide a majority of the required performance properties: good solubility, stability, wetting, oil/water interfacial reduction, emulsion stabilization, and corrosion protection.

The hydrophobic groups of the surfactants that are typically used have the equivalency of C_8 to C_{12} carbon chain lengths with alkylaryl or branched alkyl groups, providing both wetting and oily soil/water interfacial tension reduction. These molecules, due to their aryl groups or branching, also provide the required solubility and freeze–thaw stability in built detergent systems.

In alkaline cleaning systems, ethoxylated phosphate ester surfactants have excellent emulsifying properties for heavy oils and waxes. Aromatic-based phosphate esters show excellent emulsifying properties for naphthenic-based oils, whereas linear- or branched-chain ethoxylated alkyl phosphate esters provide good emulsification for paraffins and waxes (see Table 6.5).

The use of ethoxylated phosphate esters in metal cleaning processes stems also from two other of their properties: (i) their ability to act as

TABLE 6.5
Surfactants Used in Alkaline Immersion Metal Cleaning

Anionic surfactants
Ethoxylated (4–10 EO) nonylphenol phosphate esters
Ethoxylated (4–9 EO) linear (C_8–C_{10}) alcohol phosphate esters
Ethoxylated (9–12 EO) dinonylphenol phosphate esters
Nonionic surfactants
Nonylphenol ethoxylates (5–12 EO)
Octylphenol ethoxylates (5–10 EO)
Linear (C_{9-11}) alcohol ethoxylates (5–12 EO)
Branched (C_{13}–Oxo) alcohol ethoxylates (9–12 EO)
Tertiary dodecyl (branched) thioethoxylates (6–10 EO)
Amphoteric surfactants
Sodium acylamido aminopropionates $RCONHCH_2CH_2N(CH_2CH_2COO^-)_{1,2}CH_2CH_2OH$ ($R = C_{10}$–C_{14})
Sodium acylamido aminohydroxypropyl sulfonates, $RCONHCH_2CH_2N(CH_2CH(OH)CH_2SO_3^-)\ CH_2CH_2OH$ (RCO from coconut fatty acids)

hydrotropes for nonionic surfactants—i.e., to couple nonionics into built formulations, and (ii) the corrosion protection that the phosphate group in the molecule provides for most ferrous metals.

Nonionic surfactants used in these systems are typically nonylphenol-based with moles of oxyethylene ranging from 6 to 14, depending on the alkalinity in the formulation. Typically, the lower mole ethoxylates (5 to 7 EO) with an HLB of 10 to 12 provide optimum equilibrium wetting as a result of their effective oil/water and air/water interfacial tension reductions. However, for formulation compatibility and greater emulsion stability, the higher mole ethoxylates, (7 to 10 EO), with an HLB of 12 to 15, are often used. It is therefore common to find surfactant mixtures with both higher and lower HLB values formulated into these detergent systems, as this combination of surfactants will provide better cleaning and also rinsability in the subsequent rinse operation.

Acid cleaning, not as prevalent as alkaline cleaning, is used mostly to remove inorganic scale (oxidized) layers on metals prior to treatment. In most situations, the surfactants used are dodecylbenzenesulfonic acids, ethoxylated mono alkyl-phosphate esters, alkylamidosultaines, octyl/nonylphenol- or C_8 to C_{10} branched-chain alcohol ethoxylates. In this cleaning application, the soils encountered are mostly inorganic soils, and the main function of the surfactants is to provide wetting.

Spray Cleaning

In spray cleaning, the cleaning solution is circulated by pump from a stock wash solution and sprayed through a nozzle onto the substrate to be cleaned. Here, unlike in immersion cleaning, the mechanical energy imparted provides substantial cleaning power. However, the exposure of the metal to the cleaning solution is very short compared to immersion cleaning. Since the cleaning solution is sprayed under high pressure (1 to 20 kg cm^{-2}), the detergent used should be extremely low foaming and should be able to clean in a very short time.

In the selection of surfactants for spray cleaning applications, the most important interfacial property is the dynamic wetting, which correlates well with the (dynamic) reduction of surface and oil/water interfacial tension at very short times at the operating temperature (which ranges from 50 to 80°C).

Here, as in immersion cleaning, emulsification of the soils (mainly oils and waxes) by the surfactant system is important. In spray cleaning applications, it is critical to maintain a very low surface and oil/water interfacial tension not only to emulsify the contaminant soils but also to prevent redeposition of soil onto the substrate. The choice of surfactants is very much restricted by the required foaming property. The surfactant used has to produce extremely low foam to no foam. Hence, in these applications, low-foaming nonionic and amphoteric surfactants are used, and anionics, because they generally foam well, are very rarely used.

An important property for optimizing both dynamic surface and interfacial tension reduction (and therefore wetting) and foam control properties is the cloud point of the nonionic surfactant. Generally, surface activity is at a maximum and foam is at a minimum in the vicinity of the cloud point.

In order to obtain low dynamic interfacial tension, good dynamic wetting properties, and extremely low foam, nonionic and amphoteric surfactants with short linear or branched-chain hydrophobes are used. Good detergent-type nonionics, such as nonylphenol ethoxylates (7 to 12 mole EO) and linear alcohol ethoxylates (7 to 12 mole EO), foam too much to be formulated into spray cleaners.

A class of nonionics that is highly popular in spray cleaning applications is the "end capped" nonionic. These surfactants are derived from conventional ethoxylated nonionic surfactants by replacing the terminal primary hydroxyl group with either a secondary hydroxyl propyl group, a halogen, or a benzyl group. End capping of a surfactant can drastically change the physical and interfacial properties, hydrophilic–lipophilic balance, and especially the foam of the starting surfactant. In most of these applications where good detergency and low foam is essential, end capped nonionics or their mixtures are selected based on their cloud points. The insoluble monolayer formed at the aqueous/air interface at the cloud point provides minimum foam and also destabilizes any foam generated by foam stabilizing surfactants or additives. At their cloud points, these surfactants have their dynamic surface/interfacial tension reductions optimized and consequently their wetting properties as well. This is important, as the contact time of the substrate with the wash solution is very short in spray cleaning.

In general, for these "capped" nonionic ethoxylates, dynamic wetting properties are maximized when they have (C_6 to C_{12}), branched and/or cyclicle hydrophobes; oil/water emulsification properties are optimized and wetting properties are moderate when they have longer (C_{12}, C_{13}), branched hydrophobes. Although polyoxyethylene chains contribute to foaming, the moles of EO in the capped surfactant molecule are maintained at least at 60%, based on the weight of the alcohol, in order to provide good water dispersibility and O/W emulsion stability.

However, where heavy loads of organic soils are encountered, the sole use of these capped nonionics is not sufficient. Being relatively poor emulsifiers in general, these surfactants cannot hold onto the oils and greases, and very often cause redeposition of soil onto the substrate. In these situations, polyoxyethylenated (5 to 8 EO) alkyl aryl phenols or branched-chain C_{10} alcohols or thioalcohols with an HLB range of 6 to 10 are commonly used. These ethoxylated nonionics, with short-chain, branched hydrophobes and optimized EO levels, have good dynamic properties, and are excellent emulsifiers for carrying heavy soil loads. Since these surfactants have cloud point temperatures below the room temperature, they are low foaming, but not to the same extent as the corresponding capped nonionics. Therefore, in most situations capped nonionics or defoamers are added to further control the foam.

Another class of nonionic surfactants that have been developed and industrialized recently especially for Spray Cleaning Application are the alkoxylates of Nopol. These surfactants have their hydrophobes represented by a C9 bicyclic alkyl structure derived from naturally renewable raw material Pine Oil. This bicyclic structure for these molecules provides good dynamic surface tension reduction (Table 2.4), as well as wetting and extremely low foam with good rinsing properties (19). These surfactants are reported to provide extremely low ecotoxicity profile when compared to alkoxylates of linear C_{12} or branched tridecyl alcohols or nonylphenols.

In acidic or alkaline detergent formulations, the nonionics mentioned above have very poor solubility and require the use of hydrotropes (Chapter 4) to couple them into the formulation. However, in very highly alkaline and acidic formulations used in the majority of industrial applications, these nonionics show extremely poor solubility, even in the presence of hydrotropes, and are difficult to use. In these situations, zwitterionic/amphoteric surfactants with C_6 to C_{10} alkyl chains that have good alkali/acid solubility and low foam, and with effective reduction of surface tension in their aqueous solutions, are commonly used (see Table 6.6).

Pulp and Paper

Pulp Manufacture

Deresination. Deresination involves removal of resin, both saponified and unsaponified, from wood pulp in the manufacture of pulp. During

TABLE 6.6
Nonionic Surfactants Used in Spray Cleaning and Their Cloud Points, and Zwitterionic Amphoteric Surfactants Used in Spray Cleaning and Their Surface Tensions

Nonionic surfactant	Cloud point (°C)
(br)-$C_{10}H_{21}O(C_2H_4O)_{4-6}(C_3H_7O)_{7-9}H$	13–16
$C_8H_{17}C_6H_4O(C_2H_4O)_{3-5}C_6H_5$	26–28
$C_{10}H_{21}O(C_2H_4O)_{4-6}(C_3H_7O)_{4-6}H$	22–25
(br)-$C_{10}H_{21}(OC_2H_4)_{10-12}Cl$	32–34
(br)-$C_8H_{17}(OC_2H_4)_{7-9}Cl$	30–34
$C_{12}H_{25}O(C_2H_4O)_{3-5}(C_3H_7O)_{5-8}H$	23–25
$C_{12}H_{25}O(C_2H_4O)_{5-6}(C_3H_7O)_{3-4}H$	34–36
$C_{12}H_{25}O(C_2H_4O)_{7-9}(C_3H_7O)_{3-5}H$	43–45
(br)$C_{10}H_{21}O(C_2H_4O)_{4-6}H$	15–17
$C_8H_{17}C_6H_4O(C_2H_4O)_{5-7}H$	18–20
$C_9H_{17}C_6H_4O(C_2H_4O)_{5-7}H$	16–18
(br)$C_{12}H_{25}S(C_2H_4O)_{5-7}H$	15–18
Nopol $(C_3H_7O)_{3-5}(C_2H_4O)_{4-6}H$ (19)	40-45

Zwitterionic/amphoteric surfactant	γ dyn/cm (0.1% NaOH)
$RCONHCH_2CH_2CH_2N^+(CH_2CH_2COO^-)_2Na^+$ | CH_2CH_2OH	28–30
$R\,N(CH_2CH_2COO^-)_2\ 2\,Na^+$	26–28
$ROCH_2CH_2CH_2N^+(CH_3)_2CH_2CH(OH)CH_2SO_3^-$	26–30
$R\text{–}CONH\,CH_2CH_2N^+(CH_3)_2CH_2CH(OH)CH_2SO_3^-$ | CH_2CH_2OH	27–30
$RN(CH_2COOH)CH_2COO^-Na^+$	27–29
$ROCH_2CH(OH)CH_2N^+(CH_3)_2CH_2CH(OH)CH_2SO_3^-$	25–28
$R = C_6\text{–}C_{10}$	

this process, deresination agents are added to unbleached paper pulp at the hot alkali stage—at a temperature of 100 to 120°C and sodium hydroxide concentration of 4 to 8%—and cooked for one to three hours. The resin content should be reduced to below 0.3% for the manufacture of dissolved pulp. During this process hot alkali will convert the saponifiable resins to saponified fatty and rosin acids. Surfactants are used to enhance wetting by the caustic solution and the subsequent removal and emulsification of both saponified and unsaponified resins in the aqueous systems.

The major role of the surfactants in enhancing deresination of pulp is their ability to enhance the wetting and penetration of the pulp matrix by the caustic solutions. This is important, as the resins encountered are mostly water insoluble and relatively hydrophobic in nature; therefore, wetting of the resins and fiber is extremely important for the saponification as well as for the removal of the unsaponifiable resins from the fibers.

The enhanced wetting, in addition, increases the swelling of the fibers and the release of the resins from the fiber matrix. Since sufficient time exists in the deresination process for wetting and penetration to reach equilibrium, the processes are dependent on reducing the surface tension of the solution and reducing the resin/aqueous solution interfacial tension to their equilibrium values.

In addition to wetting and penetration, another important property of the surfactant is its ability to emulsify the unsaponifiable resins in the aqueous alkaline solution. This trait will help in removing the resin from the fiber in the wash stage, as well as in preventing the redeposition of the resin on the fiber. Since unsaponifiable resins (the most difficult to remove) are hydrophobic in nature, surfactants are desired that will adsorb onto the resins so as to reduce the resin/aqueous interfacial tension.

Since the deresination is carried out in hot alkaline solution, the surfactants used should show good chemical stability and solubility in the aqueous solution. Since surfactants could interfere with subsequent processing, the surfactants should have good rinsability as well.

In general, mostly ethoxylated nonionics and among anionics, ethoxylated phosphate ester surfactants, are used. Other anionics, such as sulfates and sulfonates, are not used due to their poor solubility and poor emulsifying ability in this alkaline medium.

Most of the ethoxylated nonionic and ethoxylated phosphate ester surfactants used have hydrophobes that are branched C_9 to C_{12} with aryl groups. Aryl groups, due to their polarizability, provide good affinity toward the unsaponified resins. As a result, ethoxylated nonyl- and dinonylphenol-based nonionics and phosphate esters provide excellent deresination for high resin content pulp. Table 6.7 lists surfactants commonly used in the deresination of pulp.

Paper Deinking. In the past two decades, the use of recovered pulp from wastepaper has increased tremendously because of the environmental pressure and the profitability of recycling wastepaper to make pulp. In the recovery of wastepaper, the most common contaminant to

TABLE 6.7
Surfactants Used in Deresination of Pulp

Anionic surfactants
Ethoxylated (6–10 EO) nonyl/octylphenol phosphate esters
Ethoxylated (4–8 EO) linear (C_8–C_{10}) alcohol phosphate esters
Ethoxylated (8–12 EO) dinonylphenol phosphate esters
Nonionic surfactants
Polyoxyethylene (9–15 EO) nonyl/octylphenol
Polyoxyethylene (12–20 EO) dinonylphenol
Polyoxyethylene (10–15 EO) dodecylphenol
Polyoxyethylene (10–15 EO) tridecyl (Oxo) alcohol

be removed is printing ink, as any residual ink will drastically influence the brightness of the finished paper. In addition to removing ink, the deinking process should remove other contaminants present, such as coatings, clays, resins, lattices, and adhesives. These contaminants, if not removed, can severely interfere with the papermaking process and the quality of the final paper.

The deinking of wastepaper basically involves two main processes (20): (i) pulping of the wastepaper to defiber the paper and detach the ink and other contaminants from the fiber matrix, and (ii) removal of the dispersed or suspended ink and contaminants from the pulp slurry.

A major challenge of the deinking process is the extent to which the process is "closed," to minimize the impact of the chemistry of the deinking process on subsequent steps in the papermaking process.

Pulping. The first stage in the deinking process is the pulping stage, in which a combination of chemical and mechanical treatments defiber the paper and detach the ink particles. The selection of surfactants for the pulping process will have a very marked influence on the subsequent ink removal process, whether by washing, flotation, or a "hybrid" flotation–washing process (see the following sections). Hence, choices of surfactants used at the pulping stage will depend on the subsequent ink removal process.

The Washing–Deinking Process (21). Washing–deinking is the oldest and most widely used deinking technology in North America. Washing is a mechanical process that rinses ink, ash, and dirt particles from the pulp slurry. It is most applicable for dispersible inks, mainly the carbon particles found in newsprint grades. Its efficiency is primarily a function of the size of the ink particles; it is very effective when the particle sizes have been reduced to 10 to 20 microns. Particle size reduction is therefore critical for ink removal. The particle size of the ink is reduced by treating the pulp slurry with surfactants (and alkali) that have very good wetting and dispersing properties. Through a series of dilutions, washings, and filtering processes, the dispersed ink particles are removed from the pulp slurry.

Surfactants for the Washing–Deinking Process (22, 23). As mentioned above, washing–deinking depends on dispersing the ink particles to a size small enough to be removed by rinsing through a paper fiber mat. In most washing, ink is removed as colloidal particles smaller than 20 microns.

Surfactants, by enhancing wetting of the "furnish" (wastepaper) by the aqueous alkaline solution, play a major role in washing–deinking technology. This enhanced wetting can significantly accelerate swelling of the fibers, saponification of ink resins, and release of ink particles from the fiber. Since in the pulping process sufficient time exists for wetting and penetration to reach equilibrium, processes are determined by the amount the surface tension of the solution is reduced and the amount the ink or resin/aqueous solution interfacial tension is reduced as a result of the surfactant being adsorbed under equilibrium conditions.

In addition to wetting, the surfactants should provide good dispersibility to the ink particles that are removed from the fiber. This is important, as in washing– deinking the objective is to keep the particles small enough to be removed through the washer screens. Therefore, the ideal surfactant should provide, in addition to good ink/solution interfacial tension reduction, good steric or electrostatic stability to prevent the agglomeration and redeposition of the ink particles onto the fiber. Since deinking is carried out in alkaline medium and at about 90°F (38° to 49°C), the surfactants used need to be hydrolytically stable. Anionic surfactants such as sulfonates and sulfates are rarely used due to their high foam and their poor compatibility with calcium or magnesium ions. The surfactants selected are always nonionic surfactants with good wetting and dispersing properties.

Among the nonionic surfactants most commonly used are the polyoxyethylenated alcohols or nonyl/octylphenols with at least C_9 to C_{12} alkyl chains (24). Branched alkyl chain surfactants are preferred over the linear ones, as they provide superior wetting as well as better reduction of surface and interfacial tension. In general, molecules having aromatic nucleic or olefinic bonds in the hydrophobic group are preferred, as they provide good substantivity to ink and other partially polar contaminants.

In order to enhance good dispersibility and to prevent redeposition of ink particles, polyoxyethylene/polyoxypropylene block copolymers (EO/PO/EO) are also very often used along with the ethoxylated nonionic surfactants. Among these, the most commonly used block copolymers have molecular weights in the range of 5,000 to 10,000 and have 20 to 80% by weight of ethylene oxide in the molecule. The polyoxypropylene groups in the molecule, with their multiple ether linkages, provide substantivity to the ink/resin particles, whereas the polyoxyethylene groups (end capped to minimize foaming), through their solvation, provide good steric stability to the dispersed particles.

The cloud points for the surfactants used in washing and deinking are generally 5 to 10°C higher than the pulper temperature to ensure good dispersibility of the detached ink particles (see Table 6.8).

Flotation Deinking. The flotation system incorporates a special cell that floats the ink particles to the surface in a froth that is then skimmed off. The froth is generated by introducing air, in the form of small bubbles, below the surface of the liquid in the flotation cell. Whereas washing–deinking is based on the specific size of the ink particles and their dispersibility, flotation is based on the adsorption of surfactants onto ink particles in such fashion as to cause them to be incorporated into the froth in the flotation cell. Consequently, the flotation deinking process can handle a much wider range of particle sizes, ranging from 30 to 300 microns. This feature allows a vast array of low cost wastepaper (furnish), such as varnished cardboard, cigarette cartons, copy paper, gloss printers, and so on that often contain nondispersible ink

TABLE 6.8
Surfactants Used in Washing–Deinking

Surfactant	HLB	Cloud point, °C (0.5–1.5% NaOH)	Draves wetting, sec, 40°C, 1.0% NaOH
Octylphenol ethoxylates (9 EO)	13.0	54–56	9
Octylphenol ethoxylates (11 EO)	13.5	68–72	12
Octylphenol ethoxylates (9 EO)	13.0	60–63	8
Lauryl alcohol ethoxylates (7 EO)	12.0	48–50	18
Linear (C_{9-11}) alcohol ethoxylates (6 EO)	12.5	46–48	8
Branched (C_{11-15}) secondary alcohol ethoxylates (9 EO)	13.5	56–58	9

and resinous material that cannot be handled by washing–deinking, to be deinked and deresinated efficiently. However, flotation deinking cannot handle much smaller dispersed particles (<20 microns). The type of deinking chemicals used in the flotation deinking process consequently differs from those used in washing–deinking.

Surfactants for the Flotation Deinking Process. The pulping step is similar to that in the washing–deinking process described earlier. However, the types of surfactants used in the flotation deinking step are different from those used in washing–deinking. In the flotation step, the ink dispersion in the aqueous phase must be destabilized and the particles made hydrophobic, whereas in the washing step, surfactants are designed to disperse the ink particles in the aqueous phase as small, hydrophilic colloids.

Two fundamental interfacial properties are important for these processes:

1. Reduction in the equilibrium surface tension of the aqueous solution and, reduction in the ink/aqueous interfacial tension, with consequent good equilibrium wetting by the aqueous phase in the pulping process; and
2. An increase in the ink/aqueous solution interfacial tension and a decrease in the ink/air interfacial tension with consequent poor equilibrium wetting by the aqueous phase in the flotation cell.

The surfactants used in flotation deinking very often are made to play a dual role. In the pulper, these surfactants function as wetting and dispersing agents to facilitate ink removal, similar to their function in washing–deinking; in the flotation cell, these same surfactants function as "collectors" by adsorbing onto the ink particles in a such a way as to increase the ink/aqueous solution interfacial tension and decrease the ink/air interfacial tension. The ink particles attach themselves to the small gas bubbles generated in the flotation cell due to the low interfacial tension between ink and air. The foam generated in the flotation cell carries the ink particles to the top of the flotation cell, where it is removed with the ink particles from the aqueous slurry.

In flotation deinking, use of very efficient wetting/dispersing agents is avoided, as these surfactants produce particle sizes far too small for the flotation process. More importantly, these surfactants will increase the wettability of the ink particles at the aqueous phase in the flotation stage by lowering the ink/aqueous solution interfacial tension, thus making the flotation process nonoperational.

The surfactants used as collectors in flotation systems are fatty acid soaps. Fatty acids can be added directly to the pulper machine at the alkaline pH used there. The fatty acid soaps that are used have alkyl chains preferentially ranging from 16 to 18 carbon atoms (such as stearic, oleic, palmitic, and linoleic acids). These soaps in the pulper, because they are not particularly good wetting agents, provide only limited wetting and dispersing properties in the removal of ink from the fibers. On transferring the pulp slurry to the flotation cell, the water hardness is increased by the addition of calcium chloride or carbonate. The water-soluble sodium soaps are thereby converted to their water-insoluble calcium salts, which adsorb onto the ink particles, generally with their head groups oriented toward the particles and their hydrophobic groups toward the water, making the particles hydrophobic. They consequently attach themselves to the nonpolar air bubbles and are raised to the surface of the flotation cell.

In general, long alkyl chain-based surfactants are preferred over the short or branched-chain surfactants, as the former will both adsorb more efficiently at the substrate/aqueous solution interface and also provide greater hydrophobicity to the particles. However, the use of fatty acid soaps very often results in undesirable effects elsewhere in the system. Fatty acid calcium soaps, due to their insolubility, adsorb or precipitate onto the paper fibers, resulting in greater fiber loss and also interfering with subsequent sizing operations. As a result, a tendency exists in the industry to substitute or partially supplement fatty acid soaps as the sole collectors in the flotation deinking process. This process will be discussed in the following section.

Flotation–Washing "Hybrid" Deinking. In order to improve the efficiency of ink removal from mixed wastepaper that contains newsprint (nonpolar, mainly dispersible carbon inks) as well as office waste (more polar inks that are very difficult to remove and that appear as larger particles), both washing and flotation techniques in combination are used in most North American deinking mills. These systems are referred to as flotation–washing "hybrid" deinking processes. The flotation stage is carried out at 50 to 60°C, followed by the washing step at ambient temperature, whereby the finer particles not removed by the flotation cell are removed by washing. In order to maximize the benefits of hybrid deinking, the surfactants are designed to provide both the dispersant and collector properties—very often in one type of surfactant molecule (25).

At the pulper stage, the surfactants (as in washing–deinking) should provide good wetting and dispersing properties in order to release the ink from the furnish. Depending on the furnish type, these inks are dispersed into small particles (<20 microns, to be removed at the subsequent washing stage) or as larger particles (30 to 300 microns, to be removed in the subsequent flotation stage).

Surfactants for Flotation–Washing "Hybrid" Deinking Process (26). The same type of surfactant used as the "collector" for large particles in the flotation stage is often used as the dispersant for small particles in the washing stage. This is possible because carefully selected ethoxylated nonionic surfactants are used exclusively in the hybrid deinking process. These surfactants are mixtures of materials with a distribution of polyoxyethylene chains; the component surfactants consequently have different cloud points (Chapter 4) and different tendencies to adsorb onto ink particles. The components with small polyoxyethylene chains have low cloud points and, when the temperature of the system is raised to 50 to 60°C in the flotation step, these components adsorb onto the larger, more polar ink particles *via* their polyoxyethylene chains, with their hydrophobic chains oriented toward the water. This makes the ink particles hydrophobic and thus subject to removal by flotation at this step of the process. When the temperature of the system is lowered to the ambient temperature, the surfactant components with the longer polyoxyethylene chains (which are now below their cloud points) adsorb onto the smaller, nonpolar (mainly carbon) ink particles *via* their hydrophobic groups, with their long polyoxyethylene chains acting as steric stabilizers for the dispersion of these small ink particles, which are then removed during the washing step of the process (26).

The nonionics used are based on long linear (C_{16} to C_{22}) alkyl chains, as these will provide good hydrophobicity to the ink particles in the flotation process. The majority of surfactants are ethoxylated alcohols end capped (terminated) mainly with propylene oxide, though sometimes with halogens or benzyl groups. End capping with hydrophobic groups provides good cloud point temperature control and a range applicable for the flotation process as well as low foam with small bubble sizes that will collapse easily upon reaching the surface of the flotation cell. Although alcohol ethoxylates end capped with propylene oxide provide superior performance, it is common to find EO/PO randomized alcohol alkoxylates also used in the hybrid deinking process. The advantage of using this randomized EO/PO chemistry is that these molecules have much lower viscosity and are liquids compared to the majority of the corresponding PO end-blocked long-chain alcohol ethoxylates, which are either solids or have very high viscosi-

TABLE 6.9
Surfactants Used for Flotation–Washing Deinking and Their Cloud Points (26)

Surfactant	Cloud point °C
$C_{16-20}\ O(C_2H_4O)_{10-12}\ (C_3H_7O)_{4-6}H$	55–58
$C_{16-18}\ (C_2H_4O)_{10-12}(C_3H_7O)_{4-6}H$	58–60
$C_{16-18}\ (C_2H_4O)_{8-10}(C_3H_7O)_{4-6}H$	52–54
$C_{16}\ (C_2H_4O)_{10-14}(C_3H_7O)_{5-8}H$	58–62
$C_{16}\ (C_2H_4O)_{10-14}(C_3H_7O)_{4-6}H$[a]	60–64
$C_{18}\ (C_2H_4O)_{18-20}(C_3H_7O)_{6-8}H$ (C_{18} from tallow alcohol)	56–58
RCOO $(C_2H_4O)_{6-8}(C_3H_7O)_{2-4}H$[a] (RCOO from tall oil fatty acid)	60–62
C_{18-20} C(O)O $(C_2H_4O)_{10-14}(C_3H_7O_8)_{4-8}H$	58–62

[a]Randomized EO/PO.

ties. The cloud point temperatures for the block and for the randomized copolymer with the same moles of EO and PO are within 2 to 4°C of each other, randomized EO/PO molecules giving slightly higher cloud point temperatures.

In designing the surfactants, the ratio EO/PO is selected to optimize the wetting and dispersing properties required for ink removal at the pulping and washing stage as well as to optimize the collector properties at the flotation stage. In order to provide wetting and dispersant properties, the polyoxyethylene in the molecule is maintained at a range of 50 to 60%. The majority of surfactants used have cloud points just below the 50 to 60°C flotation cell operating temperature (see Table 6.9). Studies have shown that lowering the EO contents below 50% or operating very much above the cloud point temperature, both of which decrease the solubility of the surfactant in the aqueous phase, results in poor dispersion for the inks at the washing stage and a considerable loss of fiber at the flotation process.

The benefits from using these surfactants (compared to fatty acid soaps) are increased brightness of the resultant product, insensitivity to water hardness, no scaling on equipment, low chemical carryover, and reduced fiber loss.

References

1. McWhorter, C.G., The Use of Adjuvants, in *Adjuvants for Herbicides,* Weed Science Society of America, Champaign, IL, 1982, pp. 1–8.
2. Behrens, R.W., The Physical and Chemical Properties of Surfactants and Their Effects on Formulated Herbicides, *Weeds 12*:255–258 (1964).

3. Jansen, L.L., Enhancement of Herbicide Activity: Relationship of Structure of Nonionic Surfactants to Herbicide Activity of Water-Soluble Herbicides, *J. Agric. Food Chem. 12*:223–227 (1964).
4. Manthey, F.A., Nonionic Surfactant Properties Affect Enhancement of Herbicides, in *Pesticide Formulations and Application Systems*, edited by F.R. Hall, P.D. Berger, and H.M. Collins, American Society for Testing and Materials, Philadelphia,1995, Vol. 14: ASTM STP 1234, p. 278.
5. Stevens, P.J.G., and M.J. Bukovac, Studies on Octylphenoxy Surfactants, Part 2. Effects of Folian Uptake and Translocation, *Pest. Sci. 20*:37–52 (1987).
6. Smith, L.W., C.L. Foy, and D.E. Bayer, Structure-Activity Relationships of Alkylphenol Ethylene Oxide Ether Nonionic Surfactant and Water-Soluble Herbicides, *Weed Res. 6*:233–242 (1966).
7. Tan, S., and G.D. Crabtree, Relationship of Chemical Classification and HLB of Surfactants to Upper Leaf Penetration of Growth Regulators in Apples, in *Adjuvants for Agrichemicals*, CRC Press, Boca Raton, FL, 1992, pp. 561–566.
8. Ananthapadmanabhan, K.P., E.D. Goddard, and P. Chandar, A Study of the Solution, Interfacial, and Wetting Properties of Silicone Surfactants, *Colloids and Surf. 44*:281–297 (1990).
9. Knoche, M., H. Tamura, and M.J. Bukovac, M. J., Performance and Stability of the Organosilicone Surfactant L-77: Effect of pH, Concentration, and Temperature, *J. Agric. Food Chem. 39*:202–206 (1991).
10. Murphy, G.J., G.A. Policello, and R.E. Ruckle, Formulation Consider-ations for Trisiloxane-Based Organosilicone Surfactants, in *Proceedings of the Brighton Crop Protection Conference: Weeds*, 1991, pp. 355–362.
11. Stevens, P.J.G., R.E. Gaskin, S.-O. Hong, and J.A. Zabkiewicz, J.A., Contributions of Stomatal Infiltration and Cuticular Penetration to Enhancements of Foliar Uptake by Surfactants, *Pestic. Sci. 33*:371–382 (1991).
12. Zabkiewicz, J.A., D. Coupland, and F. Ede, Effects of Surfactants on Droplet Spreading and Drying Rates in Relation to Foliar Uptake, in *Pesticide Formulations: Innovations and Developments*, edited by B. Cross and H.B.Scher, American Chemical Society, Washington, DC, 1988, ACS Symposium Series 371, pp. 77–89.
13. Policello, G.A., and S. Murphy, Surfactant Blend of a Polyalkyleneoxide Polysiloxane and an Organic Compound Having a Short-Chain Hydrophobic Moiety, U.S. Patent, 5,558,806 (1990).
14. Motooka, P., *et al.*, Gorse Control with Herbicides and Its Enhancement with Surfactants, in Proceedings of the 42nd Western Society for Weed Science Conference,1989, pp. 161–166.
15. Murphy, G.J., Policello, G.A., and Ruckle, R.E., Formulation Considerations for Trisiloxane-Based Organosilicone Adjuvants, in *Brighton Crop Protection Conference: Weeds* 4A-8, 1991, pp. 355–362.
16. Cadiergue, H., Ph.D. Thesis, Aix-Marseille III, France, 1992.
17. Gubelmann-Bonneau, I.V., P.A. Mailhe, and M.A. Perrin, Tristyrylphenol Surfactants in Agricultural Formulations: Properties and Challenges in Applications, in *Pesticide Formulations and Application Systems*, edited by F.R.

Hall, P.D. Berger, and H.M. Collins, American Society for Testing and Materials, Philadelphia, 1995, Vol. 14: ASTM STP 1234, pp. 119–136.
18. Gurin, G., and P.J. Derian, 1991, Fr. Pat. Appl. 91.02374 to Rhone Poulenc.
19. Dahanayake, M., J. Joye, and J. Le Helloco, New Generation of Nonionic Surfactants Derived from β-Piene for Industrial Applications, in *Proceedings of the 5th World Surfactant Congress*, CESIO 2000, Firenze, 2000, Vol. 2, p. 927.
20. Shrinath, A., J.T. Szewczak, and I. Bowen, A Review of Ink-Removal Techniques in Current Deinking Technology, *Tappi Journal*, July 1991, pp. 85–93.
21. Horack, R.G., and N.W. Kronlund, Principles Of Deinking Washing, *Tappi Journal*, Nov. 1980, Vol. 63 (II), 135.
22. Woodward, T.M., Appropriate Chemical Additives Are Key to Improving Deinking Operations, *Pulp and Paper, 60* (II):59 (1986).
23. Forester, W.K., Deinking of UV Cured Inks, *Tappi Journal, 70* (5):127 (1987).
24. Suwala, D.W., and Feigenbaum, H.W., A Study Of The Deinking Efficiency Of Nonionic Surfactants, in *Pulping Conference Proceedings,* Tappi Press, 1983, p. 53.
25. Dahanayake, M., GAF Corporation, *Influence Of Surfactant Structure to Performance in Hybrid Flotation—Washing Deinking*. Presented at the Tappi Conference, Pulp and Paper, 1990.
26. Jobbins, J.M., and O. Heise, Rhone-Poulenc Inc. Neutral Deinking of Mixed Office Waste in a Closed-Loop Process, Recycing Symposium, 1996, p. 27.

CHAPTER 7

Surfactant Applications 2

Construction

The construction industry is a major user of surfactants. They are widely used as water reducers (often referred to as "plasticizers"), foaming agents for concrete and gypsum, as emulsifiers for asphalt in road construction, and in the manufacture of high strength uniform glass fiber mats.

Manufacture of Uniform Glass Fiber Mats

High strength, uniform, thin sheets, or mats of glass fibers are extensively used in the building materials industry as for example in asphalt roofing shingles, backing sheets for vinyl flooring, thermal and sound insulations, and structural boards. These glass fiber mats have replaced similar sheets made traditionally of asbestos fibers. Glass fiber mats usually are made commercially by a wet-laid process (1–5).

In general, the wet-laid process for making glass fiber mats comprises first forming an aqueous suspension of short-length glass fibers under agitation in a mixing tank, then feeding the suspension through a moving screen on which the fibers enmesh themselves while the water is separated from them by suction. However, unlike natural fibers, such as cellulose or asbestos, glass fibers do not disperse well in water. When glass fibers, which come as strands or bundles of parallel fibers, are put into water and stirred, they do not form a well-dispersed system. In fact, upon extended agitation, the fibers agglomerate as large clumps which are very difficult to redisperse. In order to overcome this inherent problem with glass fibers, it has been the practice in the industry to use dispersing aids for the glass fibers, in order to keep the fibers separated from one another and dispersed in the aqueous phase. To produce high strength, uniform thin sheets or mats, it is essential to have the fibers fully dispersed in the aqueous phase prior to settling on the moving screen. Also, uniformity in distribution and separation of individual fibers is important so that a uniform distribution of the resin or other additive material throughout the mat can be achieved, thereby preventing channels for water flow developing at areas of weakness now covered by the resin.

Dispersibility and wettability are the most important interfacial properties that are required in the selection of surfactants for the manufacture of glass fiber mats. Since glass fibers are slightly negatively charged, the majority of surfactants used are amine-based surfactants that are capable of adsorbing through electrostatic interaction onto the negatively charged sites of the glass fibers. Among the amine-based surfactants, alkylamine oxides and polyoxyethylenated amines that are either quarterized or very easily protonated are commonly used. Once they are adsorbed, the negatively charged oxygen atom in amine oxides and the polyoxyethylenated chains in the case of the ethoxylated amines provided good wetting and steric and/or electrostatic stability for the dispersions of the glass fibers.

In order to promote good substantivity and close packing for the surfactant molecules, the hydrophobic groups of the surfactants that are typically used have the equivalent of C_{12} to C_{18} carbon chain lengths, with linear alkyl groups being preferred. In situations where high glass fiber loading mats are manufactured, it is common to find C_{16} to C_{18} alkylated amine oxides or amine alkoxylates (6–8), as their longer chains prevent abrasion and damage to the fibers on close contact. In the case of the polyoxyethylenated amines, in order to provide good steric stability, higher amounts of ethoxylation, typically varying from 9 to 15 moles, are generally used. In order to control foaming in these fast turn over wet-laid

TABLE 7.1
Surfactants Used for Glass Fiber Dispersion in Making Uniform Glass Fiber Mats (6–10)

$R\ N^{+}(CH_3)_2O^{-}$
$R_1\ N^{+}(CH_2CH_2OH)_2O^{-}$
$R\ N\ [(CH_2CH_2O)_mH]_n$
$R\ N\ [(C_3H7O)_x\ (C_2H_4O)_yH]_2$
$R\ N\ [(C_3H7O)_x\ (C_2H_4O)_yH]_2$
$R\ N^{+}\ (CH_3)_2\ CH\ CH\ (OH)\ CH_2\ SO_3^{-}$
$R = C_{16}$ to C_{18}
$R = C_{15}$ CO NH $CH_2CH_2^{-}$ to C_7 CO NH $CH_2CH_2^{-}$
m = 9 to 15, n = 2
m = 8 to 10, n =3
x = 12 to 15, y = 2 to 4

processes, in recent years, popularity of propylene oxide capped amine ethoxylates are growing significantly (7, 9–10).

Concrete

In the concrete industry, the concretes produced either by precast or and in ready-mixed form use what are termed "admixtures," which contain as their major ingredient one or more surface-active agents. Admixtures used in the concrete industry are broadly classified into water-reducing and air-entraining admixtures. The purpose of the water-reducing admixtures (plasticizers) is to improve the workability (i.e., reduce the viscosity) of the cement slurry at reduced water/cement ratios. Without the plasticizers for good handling and workability, cement requires twice as much water. This additional water causes delays in setting the cement and produces voids in the concrete, with consequent severe losses of strength. The purpose of the air-entraining agents is to introduce a controlled amount of air into the concrete. This air entrainment increases the freeze–thaw resistance of the concrete. It also decreases its density and improves its workability.

The surfactants used in the concrete industry as plasticizers must be compatible and stable in highly alkaline aqueous systems. Furthermore, aqueous cement mixes are high in many metal ions (Al, Fe, Ca, Si). Therefore, the surfactants used must tolerate many metal ions and remain effective in their presence.

Due to the high loading of cement in a concrete mix slurry and to incomplete hydration, cement grains very often form aggregates in water.

The majority of surfactants used in the industry for improving the workability of cement are anionics with a high charge density, i.e., sulfates and sulfonates having short alkyl chains. These surfactants in highly alkaline and electrolyte-rich aqueous solutions provide low solid/liquid interfacial tension as well as good electrostatic stabilization of dispersed cement particles. The low solid/aqueous solution interfacial tension produced when these surfactants adsorb onto the solid particles (with their charged groups facing the water) provides good wettability, which enhances the homogeneity and uniformity of hydration of the particles during the reaction of the cement particles with water. The high charge density imparted to the particles as a result of the surfactant adsorption at the solid/liquid interface provides not only good dispersibility in water but also results in much lower viscosity at a given water/cement ratio (due to particle–particle repulsion), good workability, and ultimately much greater strength in the concrete.

Use of surfactants with good wetting and dispersing properties can reduce the water requirement by 10 to 15%. Maintaining the same workability of the concrete mix at a higher cement/water ratio allows much easier and faster setting and compacting of the concrete on site and ultimately greater strength of the concrete.

The most popularly used surfactants for water reducers are the ligno- and naphthalenesulfonates. Short or branched alkyl chains and aryl groups are preferred over long chains due to their better solubility in highly alkaline and high electrolyte-rich aqueous solutions. Also, surfactants that have multiple anionic groups in the molecule, such as naphthalene sulfonic acid–formaldehyde condensates, are frequently used. The high charge density produced on the surface of the cement particles upon adsorption gives the particles excellent dispersibility and gives stability to the resulting cement dispersion. Naphthalene sulfonic acid–formaldehyde and also sulfonated melamine-formaldehyde condensates that have approximately six sulfonic acid groups per molecule are capable of reducing the water required by 15 to 25%, and are referred to as "superplasticizers." Due to their excellent dispersing properties and lower retardation of concrete hardening compared to lignosulfonates, these dispersants are used for the preparation of high-strength concrete. Superplasticizers are very often used in the formulation of flowing concrete, i.e., concrete having high plasticity with good flow at a low water/cement ratio. Again, this is due to their exceptional dispersing properties, which produce much lower viscosities and high plasticity at these ratios.

In the manufacture of freeze-thaw resistant concrete, surfactants are used to provide controlled amounts of well-dispersed, uniform air bubbles in the cement. In this case, the preferred surfactants are those that provide good interfacial tension reduction at the air/concrete interface in solutions of high alkalinity and electrolyte content. For this application, sulfonates and sulfates with relatively short alkyl chains (C_8 to C_{10}) or branched C_{12} alkyl chains, or rosin acid soaps are used. The alkyl sulfonates and sulfates, compared to their long-chain analog, are more tolerant to both alkalinity and electrolyte, and lower the tension at the air/aqueous interface. The rosin soaps adsorb onto the surface of the cement particles via their carboxylate groups and render the cement surfaces hydrophobic, thus reducing the air/cement interfacial tension. Table 7.2 lists the surfactants used in concrete production and their major functions.

Gypsum Board

In the production of gypsum board, surfactants are used mainly for air entrainment (foaming), but also for plasticizing (reducing water). Foam plays a major role in reducing the density of the board, which allows for ease of handling, thermal insulation, soundproofing, shorter drying times, and cost savings in raw materials.

Compared to water reducing surfactants used in cement mixtures, the preferred surfactants are those that have exceptionally good foaming in air/aqueous solutions that are highly alkaline and contain heavy metal ions. In addition, the surfactants must provide good dispersing properties that lower the water/gypsum ratio (see Table 7.3).

The surfactants that are most widely used are the sulfated anionics with a high charge density, with short alkyl chains ranging from C_6 to C_{11}. These surfactants produce very low gypsum/aqueous solution and air/aqueous solution interfacial tensions in the presence of electrolytes and

TABLE 7.2
Surfactants Used in Concrete—Major Functions

Surfactant	Property
Lignosulfonates	Dispersing
Sodium butyl or isopropyl naphthalene sulfonates	Wetting
Sodium naphthalene sulfonic acid–formaldehyde condensates	Dispersing
Sodium alkyl (branched C_8–C_{10}) sulfates	Wetting and dispersing
Sodium alkyl (C_6–C_{10}) ethoxy (2–4) sulfates	Foaming and air entrainment
Rosin acid soaps	Foaming and air entrainment

TABLE 7.3
Surfactants Used in Gypsum Board Manufacture

Surfactant	Waring blender foam ht, cm in 4% brine
$R\,O(C_2\,H_4O)_{2-4}SO_4^-\,Na^+$	14–16
$R\,(OC_2H_4)_{4-6}\,O\,P(O)\,(OH)_2$ Mono/di 90:10 $R = C_6–C_{11}$	12–16

high alkalinity, thereby having excellent foaming ability. Their high charge density gives good dispersibility to the particles at a low water/gypsum ratio. The longer-chain anionic ($>C_{10}$) surfactants show only a limited solubility in the medium and are rarely used. In addition to sulfates and sulfonates, short chains C_8 to C_{11}, ethoxylated phosphate ester surfactants high in mono alkylates are sometimes used. Branched alkylated anionics are rarely used, as they provide relatively unstable foams that tend to leave voids when the gypsum sets.

Asphalt Emulsions

Asphalt (bitumen) used in road construction is a solid at room temperature and is nonpolar. In order to permit handling the asphalt in a fluid form, the asphalt is used in the form of an asphalt/water emulsion. To emulsify the asphalt and also to improve its wetting of and adhesion onto polar substrates, specific surfactants are used in the formulation of these emulsions.

The surfactants used in the formulations of asphalt emulsions for road construction must play a dual role. They must first reduce the asphalt/water interfacial tension so that the asphalt can be emulsified in the water. However, when the emulsion contacts the road-building material ("aggregate"), the emulsion must wet the aggregate, and the surfactant in it must preferably adsorb onto the aggregate in such fashion as to render it hydrophobic, thereby promoting adhesion to it of the nonpolar asphalt. In addition, this preferred adsorption of the emulsifying surfactant onto the aggregate should cause the emulsion to "break," with consequent deposition of the asphalt onto the now hydrophobic aggregate surface.

Since road-building aggregate is usually negatively charged, the majority of surfactants used as emulsifiers are cationic surfactants with C_{12} to C_{20} alkyl chains. The cationic surfactants used are either quaternized cationics or alkylated poly-amines that are easily protonated in the presence of strong inorganic acids. These surfactants, with their long alkyl chains, produce sufficiently low interfacial tensions at the asphalt/water interface to produce emulsions that are stable during storage and transportation. Due to their cationic nature, these surfactants adsorb strongly onto negatively-charged surfaces with their hydrophilic head groups oriented toward the surface and their long hydrophobic groups turned toward the water. This increases the substrate/water interfacial tension and also decreases the nonpolar bitumen/substrate interfacial tension. Depletion of the emulsifier as a result of either adsorption onto the substrate or loss of surface activity through neutralization by the calcium carbonate (in the case of the protonated alkylpolyamines), breaks the emulsion and facilitates setting of the bitumen onto the substrate. The concentration of the emulsifier and the type of emulsifier used generally control the setting time for the asphalt. In the case of rapid-setting emulsions, which are used for spraying onto new surfaces or as seal coats, high charge density alkyl polyamines or alkyl amidoamines with hydrochloric acid are commonly used. They are used at low concentrations (0.2 to 0.3% by weight) of the bitumen. Their rapid deprotonation on contact with calcium carbonate produces rapid demulsification or "breaking" of the bitumen/water emulsion.

In the case of medium- to slower-setting bitumen emulsions, which are used in paving or slurry sealing, much higher concentrations of emulsifiers are used to provide greater emulsion stability. In these situations, it is more common to use ethoxylated amines or quaternized cationics that provide better emulsifying power than the polyamines. In general, quaternized emulsifiers provide good workability and strong adhesion and antistripping properties for asphalt, even when sprayed onto wet surfaces. In extremely slow-setting emulsions, used in situations in which it takes a long time for the asphalt emulsion to be transported to the construction site, nonionic surfactants are very often used, in addition to cationic surfactants, to provide long-term stability to the emulsions.

Ethoxylated nonionic surfactants with branched alkyl chains are used for extremely "slower setting" bitumen emulsions, or for cement asphalts in which the acidity of the cement prevents the use of cationic surfactants. Also, where freeze–thaw stability or low viscosity of the emulsion is desired, these surfactants are used along with the cationic surfactants. Table 7.4 gives the surfactants commonly associated with asphalt emulsions.

TABLE 7.4
Surfactants Used in Asphalt Emulsions

Cationic surfactants
$RCONHCH_2CH_2NH_3^+X^-$
$R\ N^+H_2CH_2CH_2\ NH_3^+\ 2X^-$
$R\ N^+H_2CH_2CH_2CH_2NH_3^+2X^-$
$R\ N^+(CH_3)_3Cl^-$
$R\ CONHCH_2CH_2\ N(CH_3)_3^+X^-$
$R\ N(CH_2CH_2OH)_2$
$R\ N^+(O^-)(CH_3)_2$
$R = C_4$ to C_{18}
$X^- = Cl^-, Br^-, CH_3SO_4^-$

Oil Fields

Aqueous Fracturing Fluids

In recovering petroleum from subterranean formations, it is common practice to fracture or crack the rock containing the petroleum in order to create flow channels. Viscous aqueous or nonaqueous fluids are hydraulically injected into the well bore to accomplish this function.

Once the rock is fractured, a "proppant" generally consisting of sand or gravel that is suspended in the viscous fracturing fluid is placed in the fracture. These proppants prevent collapse of the fracture and provide improved flow of oil or gas into the well bore. More importantly, in situations in which the well bore traverses a water-bearing zone before penetrating the petroleum-bearing zone, the viscous fluids play an important role. Here they are able to block or plug the pore structure of the water-bearing zone (due to their high viscosity) and prevent formation water from seeping into the well bore once the well is in production. Traditionally, the fracturing fluids used have been hydratable polysaccharides such as galactomannoses (guars) cross-linked with zirconium, beryllium, etc. as viscosifiers. Several problems have been associated with the use of polysaccharides together with highly elastic modules, such as guars, as viscosifiers for fracturing fluids: (i) the process of hydration and cross-linking with metal ions is cumbersome, takes time to react, and requires labor and expensive equipment; (ii) controlling viscosity with the cross-linking agent is very concentration dependent; and (iii) recovering the polysaccharides, once the fracturing operation is completed, is very tedious and involves oxidation or enzymatic hydrolysis of the hydrocolloid; even then, the recovery of the degraded polymer is rarely complete.

In order to overcome the above-mentioned limitations with hydrocolloids, the use of surfactants to provide the viscosity and the viscoelasticity properties is becoming increasingly popular in fracturing fluids (11, 12).

The surfactants required for use in fracturing fluids are those that, when used at a low concentration (<5%), give viscoelastic solutions—viscous liquids having the capacity to return to the original state or form when an applied stress is released. Viscoelasticity ensures good pumpability and handling under high shear rate and viscosity buildup under low shear rate. This property is important for fracturing and suspending proppants, and for plugging flow channels in the water-rich zones.

As fracturing is carried out at high pressures and at elevated temperatures (>200°F), the surfactants used should be chemically stable and show minimum fluctuation in viscosity with temperature and pressure. Ability to achieve these properties at low concentrations is important for the following reasons: (i) to be cost effective, (ii) to prevent contamination of the well bore and reservoirs, (iii) to prevent wetting of clay, and (iv) to avoid emulsification of petroleum and water. Furthermore, low concentration usage results in a rapid drop in viscosity when in contact with formation water, therefore allowing the recovery of surfactant once the fracturing operation is completed.

Surfactants that are now being used and that have shown very good results are the long-chain, linear, alkylated C_{18} to C_{22} cationic surfactants capable of forming "worm-like" micelles (as opposed to the spherical micelles formed by most other surfactants) in the presence of organic and inorganic salts such as KCl, NH_4Cl, and salicylates (Chapter 3, Viscosity of Micellar Solutions). These worm-like micelles are responsible for providing good viscoelasticity. In contrast to other long-chain types of surfactants, this class of surfactant is capable of forming these association structures at concentrations of less than 5% (by wt). Furthermore, these association structures are formed in a dynamic state and under shear pressure, and in the presence of hydrocarbons, the structures are readily altered. Under the latter conditions, the micelles rapidly revert to spherical micellar or monomer forms that do not build viscosity in the fracturing liquid. In general, the longer the hydrocarbon chain (C_{20} to C_{22}), the higher the viscosity and the lower the sensitivity of the viscosity to the temperatures important in oil field applications (>200°F) (see Table 7.5).

TABLE 7.5
Surfactants Used in Aqueous Fracturing Fluids

Surfactant	Viscosity 100^{-1}s/shear rate at 5% (by wt)
$C_{16}H_{33}N^+(CH_3)_3 \cdot CH_2(COO^-)_2$	87
$C_{18}H_{37}$ $N^+(CH_3)_3 \cdot HOCH_2C_6H_4COO^-$	90
$C_{18}H_{37}$ $N^+(CH_3)_3 \cdot CH_2(COO^-)_2$	110
$RN^+(CH_3)_3 \cdot CH_2(COO^-)_2$	140
$RN^+(CH_2CH_2OH)$ $_2CH_3Cl^-$	180
R = C_{22}, rapeseed oil-derived erucyl	

These cationics also have good solubility in hydrocarbons. Therefore, when they reach the petroleum zone and come in contact with hydrocarbons, depletion of the surfactants from the aqueous phase takes place and this results in their reverting from a worm-like micellar structure to a spherical or monomeric form with much reduced viscosity. This prevents plugging of the hydrocarbon flow channels by the fracture fluid while maintaining the plugging of the water flow channels.

Firefighting Foams

Foams play an important role in extinguishing and preventing oil fires. Essentially, they reduce the density of water relative to that of the oil or gasoline, allowing an aqueous solution in the form of a foam to spread over the fuel, thereby preventing any volatile fuels from coming into contact with the oxygen present in air.

Although many types of foaming agents are designed for fighting various types of fires, the most sophisticated and important are the foaming agents used for fighting the oil or gasoline (hydrocarbon fuel) fires that are the most prevalent (second only to forest fires).

In the formulation of foams for firefighting, surfactants are a key ingredient in providing foaming capacity to aqueous solutions. For use in the formulation of firefighting foams, surfactants must have three properties:

1. They must cause wetting or spreading of the aqueous foaming solution over the fuel (to form a barrier against the escape of fuel vapor).
2. They must show excellent foaming power and foam stability in hard and salt water.
3. They must show poor oil/water emulsification.

A clear relationship can be found between the enhanced wetting or spreading of the aqueous foam solution over the fuel and its effectiveness in extinguishing the fuel fire. Since sufficient time exists for spreading or wetting to reach equilibrium, the process is determined by the equilibrium spreading coefficient (Eq. 2.8); consequently, the desired surfactant properties are reduction of surface tension and reduction of the fuel/aqueous solution interfacial tensions. However, reduction of the oil/solution interfacial tension can lead to emulsification of the oil in water, with consequent spreading of the fire. As a result, the surfactants used for this application are those that will only moderately lower the oil/water interfacial tension (to not less than 1 dyn cm^{-1}). In addition, the surfactants used must be very good foaming agents, must be stable up to high temperatures, and must show good solubility in hard and salt water. Since many good foaming agents are also good emulsifiers for oil/water systems, the surfactants must be specially selected to have relatively poor emulsifying properties.

Formulations for oil and gasoline firefighting foams are generally made from blends of surfactants, in order to satisfy all the requirements needed for performance (13–14). The main surfactant used, however, is a short-chain C_6 to C_{10} fluorosurfactant, either zwitterionic or anionic. These surfactants have the capacity to reduce the surface tensions in aqueous solutions to 15 to 20 dyn/cm and have relatively high spreading rates. Since fluorinated surfactants are both hydrophobic and oleophobic, they are not good at lowering the oil/water interfacial tension. In order to provide good wetting, the spreading coefficient for the formulation should be positive (Eq. 2.8). Since the surface energy of the oil/gasoline fuel is in the range of 20 to 25 dyn/cm and the fuel oil/water interfacial tension is at least 40 to 50 dyn/cm for the spreading coefficient to be positive, the interfacial tensions of fuel oil/water should be reduced to at least below 5 dyn/cm (see Table 7.6). This is very often accomplished by using a C_6 to C_{10} or alkylated anionic surfactant, preferably a sulfate or sulfonate (15–17). These surfactants, in combination with fluorocarbon surfactants, are synergistic in lowering the oil/water interfacial tensions below 5 dyn/cm, which is sufficient to make the spreading coefficients positive. This allows good spreading of hard water on the flammable liquids. Also, the use of these short-chain alkylated anionics assures poor emulsification and emulsion stability for the oil/water systems (see Table 7.7).

TABLE 7.6
Fluorosurfactants Used in Hydrocarbon Firefighting Foams

Fluorosurfactants	γ, dyn/cm⁻, 0.1% (by wt)	Ross-Miles foam ht, 0.1% (by wt)	γ_{int},[a] dyn/cm, 0.1% (by wt)
$C_{6-10}F_{13-21}CH_2CH(OCO\ CH_3)\ CH_2\ N^+(CH_3)_2\ CH_2COO^-$	18.2	220	7.2
$C_8H_{17}CH_2CH_2\ S\ CH_2CH_2CONH\ C(CH_3)_2\ CH_2SO_3^-Na^+$	16.2	—	—
$C_{6-10}F_{13-21}CH_2CH_2N^+\ (CH_3)_3\ CH_3SO_4^-$	19.2	210	6.8
$C_{6-10}F_{13-21}SO_2N(CH_2\ CH_2)\ C_3H_6N^+(CH_3)_3 \cdot CH_3SO_4^-$	18.5	215	7.5
$C_2F_4CONH\ C_3\ H_6N^+(CH_3)_2\ CH_2\ CH_2\ CO_2^-$	17.2	225	8.0
$C_8F_{17}CH_2CH_2\ S\ CH_2\ CH_2\ COO^-Li^+$	17.1	180	6.5

[a]Abbreviation: γ_{int}, heptane/water.

Textiles

Antistatic Agents in Spin Finish Formulations

Surfactants play a major role in the formulation of spin finishes, coatings used in the spinning of textile fibers. Although spin finish formulations provide a multitude of performance properties, this discussion will focus on the role played by surfactants in providing antistatic properties for the textile fibers.

TABLE 7.7
Interfacial and Spreading Coefficients for Some Commercial Surfactants and Their Blends

Surfactant system	C (% by wt)	γ (dyn cm⁻¹)	γ_1 (dyn cm⁻¹)	S_{CH}[a]
$C_{6-10}F_{13-21}CH_2CH(OCOCH_3)\ CH_2\ N^+(CH_3)_2CH_2COO^-$	0.09	16.3	7.4	1.3
$CH_3(CH_2)_3CH(C_2H_5)CH_2\ SO_4^-Na^+$	0.09	40.2	23.7	—
$C_{6-10}F_{13-21}CH_2CH(OCOCH_3)\ CH_2\ N^+(CH_3)_2CH_2COO^-/CH_3(CH_2)_3CH(C_2H_5)CH_2\ SO_4^-Na^+$	0.09	15.5	4.9	4.6

[a]Abbreviation: S_{CH}, spreading coefficient on cyclohexane.

During fiber spinning, drawing, and texturing, the synthetic fibers (nylon, polyester, polyamides, etc.) undergo considerable friction, which produces a buildup of static charges on the fibers, resulting in poor spinnability due to fluffing and roller wrapping of the fibers. To overcome this problem and to increase the spinning efficiency of the yarn, antistatic agents, more particularly surfactants that provide good static control, are always used.

In order to provide good antistatic properties, the surfactants selected must have good adsorption onto the synthetic fibers and must adsorb in such fashion as to form a hydrophilic conductive ionic film on the surface of the fiber. The adsorbed film should facilitate the absorption of a film of moisture from the atmosphere to enhance surface conductivity, thereby providing static control of the fibers. The selection of surfactants as antistatic agents is also based on their ability to provide good lubricity and good emulsifying properties to the lubricants (mineral and vegetable oils) that are often used in spin finish formulations.

The most widely used antistatic agents in spin finish formulations are anionic surfactants. Ethoxylated nonionic surfactants, which are not as good as anionic surfactants in controlling static buildup, are used in situations in which performance criteria are much lower, such as for slower spin speeds where static charge buildup is not excessive.

Among anionic surfactants, phosphate ester surfactants with linear C_8 to C_{16} alkyl chains that produce very low fiber/liquid interfacial tensions and also form closely packed monomolecular layers that result in conductive films on the fiber surface, are widely used. Ethoxylated phosphate ester surfactants with C_{12} to C_{18} alkyl chains provide excellent fiber/fiber and fiber/spindle lubrication as well as good emulsifying properties for the lubricants used in spin finish formulations. Although higher alkyl chain lengths (>12 carbon atoms) help in lubricity and emulsification, the alkyl chains for the anionic surfactants used in spin finish formulations are usually restricted to C_{10} or C_{14} in order to optimize the antistatic properties. This is because it has been observed that static control decreases as the alkyl chain lengths are increased beyond C_{14}.

Shorter alkyl chains (<10 carbon atoms), although they provide good antistatic properties, provide poor lubricity and emulsification properties. Also, for a given phosphate ester surfactant, monoester, $R(OC_2H_4)_x OP(O)(O^-)_2$ to diester, $[R(OC_2H_4)_x O]_2 P(O) O^-$, ratios >50% provide better static control due to the higher charge density of the monoester compared to the diester. Incorporating oxyethylene groups in the molecule further contributes to good conductivity due to their tendency to absorb and retain moisture from the atmosphere (see Tables 7.8 and 7.9).

Industrial Water Treatment

Synthetic water soluble polyelectrolytes are used extensively in the treatment and purification of industrial wastewater. Their mode of action is through electrostatic interaction with oppositely-charged suspended waste solids in the effluent water. These interactions lead to a decrease of the electrical charges on the suspended particles and result in the particles coalescing into a mass large enough to be separated from the aqueous medium. Since most of the solid waste particles found in industrial effluents are negatively charged, the polyelectrolytes used in wastewater treatment are positively charged. In this respect, the most frequently used polymers are copolymers of amine monomers, such as methacrylamidopropyltrimethylamine, acrylamidopropyltrimethylamine, acryloyloxyhydroxypropyltrimethylamine, etc.

These polymers are made by emulsion polymerization technology, in which the monomers are polymerized in water/paraffinic oil emulsion systems. The type of surfactants and the ratio of water to oil is selected so that the final product is a water-in-oil emulsion with the final polymer finely dispersed in the internal aqueous phase. This technique allows the synthesis of very fine dispersions with polymer particle sizes in the range of 0.1 to 10 mm, and polymer concentrations of 30% or greater without any increase in the viscosity of the medium.

TABLE 7.8
Surfactants Used as Antistatic Agents in Spin Finish Formulations

Anionic surfactants
Ethoxylated (6–10 EO) dodecylphenol phosphate esters
Ethoxylated (5–10 EO) linear (C_{10}–C_{16}) alcohol phosphate esters
Ethoxylated (5–10 EO) tridecylalcohol phosphate esters
Ethoxylated (0–4 EO) alkyl (C_{12}–C_{14}) sulfates
Nonionic surfactants
Ethoxylated (15–20 SEO) castor oil sorbitan monolaurate
Ethoxylated (5–10 SEO) sorbitan monolaurate
Cationic surfactant
Ethoxylated (6–12 mol SEO) tallow amines

TABLE 7.9
Resistivity Measurements of Some Phosphate Ester Surfactants on Polyester Filament at 0.3% Concentration (by wt)

Surfactant	Resistivity log ohm/cm
Control (no surfactant)	20.50
C_{8-10} ethoxylated (5 EO) phosphate esters Mono/di, 60:40 (by wt)	10.50
C_{10} ethoxylated (6 EO) phosphate esters Mono/di, 60:40 (by wt)	10.75
C_{12} ethoxylated (4 EO) phosphate esters Mono/di, 80:20 (by wt)	11.25
C_{12} ethoxylated (6 EO) phosphate esters Mono/di, 60:40 (by wt)	10.90
C_{12} ethoxylated (4 EO) phosphate esters Mono/di, 90:10 (by wt)	10.70
C_{12} phosphate esters Mono/di, 80:20 (by wt)	11.80
C_{14-16} ethoxylated (5 EO) phosphate esters Mono/di, 60:40 (by wt)	12.20
C_{16} ethoxylated (5 EO) phosphate esters Mono/di, 60:40 (by wt)	13.25
C_{18} ethoxylated (6 EO) phosphate esters Mono/di, 60:40 (by wt)	13.20

When these W/O emulsions are used in the treatment of wastewater, a key step is releasing the polymer from the internal aqueous phase of the emulsion to the external aqueous medium that is being treated. Here, in order to be effective, the polymer must be released as quickly as possible as a fine dispersion in order for the polymer to make intimate contact with the oppositely-charged suspended solids. This process of releasing the polymer is achieved efficiently by judiciously selecting surfactants that are added to the W/O emulsion just prior to the application (18–20). These surfactants facilitate release of the polymer from the internal aqueous phase of the water-in-oil emulsion to the external aqueous medium by inverting the water-in-oil emulsion to an oil-in-water emulsion. Surfactants that effectively accomplish the process rapidly are referred to as "inverting" surfactants.

In selecting the surfactants, the most important interfacial properties that influence the inversion process are the rapid (dynamic) reduction of surface and O/W interfacial tensions and the tendency of these surfactants to form oil-in-water emulsions. However, the choice of surfactants used for water treatment is limited to nonionic surfactants, as very often ionic surfactants interfere with the interactions of polymer and the suspended solids. In addition to providing low O/W interfacial tensions rapidly, the surfactants used have to provide sufficient stability to the resulting oil-in-water emulsion. This trait is important in order to prevent the coalescence of oil particles and any entrapment of the polymer by the oil. In addition, the surfactants selected must enhance the interaction of the polymer with the solid waste particles. This is normally accomplished by the decrease in surface tension at the aqueous phase, resulting in increased wettability of the suspended solids.

In order to provide the desired interfacial properties (low interfacial tensions, good emulsion stability, and wetting), nonionic surfactants with branched hydrophobes are used extensively in these applications. Among the surfactants popularly used are the ethoxylated nonyl, octyl, dinonylphenols, and ethoxylated tridecyl alcohols with typical HLBs in the range of 8 to 12. The branched hydrophobes provide much lower dynamic interfacial tensions than the linear analogs. Further, these surfactants, with their long polyoxyethylene groups, provide steric stability for the oil-in-water emulsions as well as good dispersion for the polymers as soon as they are released to the external aqueous medium. The latter property prevents the "local" buildup of concentrations of the polymer, its aggregation, or any viscosity buildup that is detrimental to the performance of the polymer. Inverting surfactants commonly used for water treatment applications are listed in Table 7.10. Inversion times for some of the surfactants are shown in Table 7.11.

Metalworking

Metalworking fluids (MWF) are used to provide lubricity and cooling during the many metal-cutting operations. Surfactants play a major role in providing the formulation of MWF with a multitude of properties. They are used in MWF as emulsifiers, lubricants, dispersants, wetting agents, or as corrosion inhibitors.

MWF are classified into four major categories based on their formulation compositions: straight oils, soluble oils, semi-synthetics, and synthetics.

1. *Straight oils.* These are the oldest type of MWF used in the industry. The formulations are simple, based 90% or greater on petroleum oils, and are used directly without dilution. In these formulations, the petroleum oils provide the bulk of the lubrication and cooling. Surfactants are used to enhance lubricity and wetting properties of the base oils.

TABLE 7.10
"Inverting" Surfactants Used with Water Treatment Polymers

(br)$C_9H_{19}C_6H_4O(C_2H_4O)_{8-12}H$
(br)$C_8H_{17}C_6H_4O(C_2H_4O)_{7-10}H$
(br)$C_{12}H_{25}O(C_2H_4O)_{8-12}H$
(br)$C_{13}H_{27}O(C_2H_4O)_{8-10}H$
$C_{18}H_{35}O(C_2H_4O)_{15-20}H$
Castor oil $(C_2H_4O)_{20-30}H$

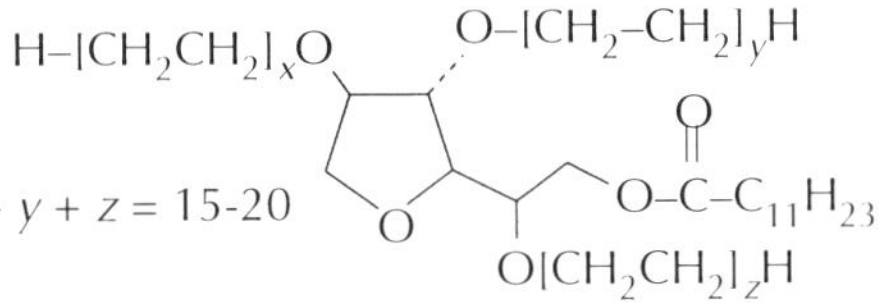

$x + y + z = 15\text{-}20$

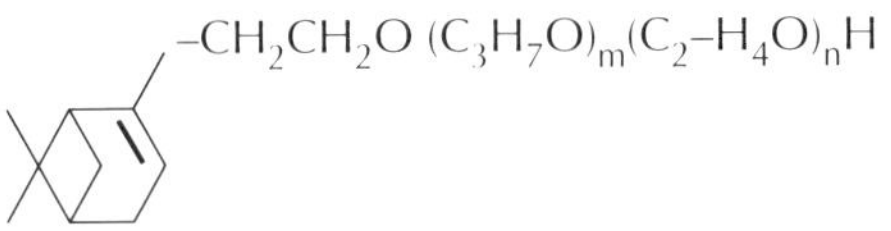

m = 3 to 4; n = 6 to 8.

2. *Soluble oils.* These formulations are similar to straight oils and are based on petroleum oils, but they are diluted with water to form an oil-in-water emulsion prior to use or at the time of application. Surfactants provide lubricity and emulsification for the oils on dilution with water, prior to use.

TABLE 7.11
"Inversion" Time for Some Commercial Surfactants

Surfactant	"Inversion" time (sec)[a]
Control (no surfactants)	>300
(br)$C_8H_{17}\,C_6H_4\,O(C_2H_4O)_8H$	13
(br)$C_8H_{17}\,C_6H_4\,O(C_2H_4O)_9H$	15
(br)$C_9H_{19}\,C_6H_4\,O(C_2H_4O)_9H$	18
$(C_9H_{19})_2C_6H_3\,O(C_2H_4O)_{15}H$	17
(br)$C_9H_{19}C(CH_3)_2\,S(OC_2H_4)_9OH$	14
$C_{12}H_{25}\,O(C_2H_4O)_9H$	25
Castor oil $O(C_2H_4O)_{30}H$	18

[a]Time taken for 0.5% (by wt) of the water-soluble polymer (in W/O emulsion) to reach its maximum viscosity at ambient temperature. This polymer is a copolymer of acrylamide and acryloxyhydroxyethyl trimethylammonium chloride in W/O emulsion.

3. *Semi-synthetic oils.* The majority of these MWFs are formulated as mineral oil in water microemulsions. These formulations are easily converted to O/W macro-emulsions on dilution with water, at the time of application. Surfactants are used primarily as emulsifiers in the formulation of the microemulsion concentrates, as well as in the subsequent emulsification of the oil phase on dilution with water.
4. *Synthetics.* These fluids are petroleum- or mineral-oil free and are for the most part water-based. Several additives, including surfactants, are incorporated to provide the formulation with lubricity, rust inhibition, and defoaming properties.

Lubricity and emulsification are the most important interfacial properties that are required from surfactants in formulating MWF. When used for providing lubricity, surfactants must have the ability to adsorb onto the metal with the orientation of their hydrophobes away from the metal surface to form a tightly packed monomolecular film. The hydrophobic film must provide lubricity by reducing friction between the cutting or grinding metal surfaces. In addition to reducing friction, the monomolecular film should provide good metal/petroleum oil interfacial tension reduction in order to enhance the wetting of the substrate by the hydrocarbon oils present in straight oils, soluble oils, and semi-synthetic MWF.

In the formulation of both soluble and semi-synthetic MWF, the surfactants should provide, in addition, good spontaneous emulsification for the MWF upon dilution with water, providing a fine dispersion of oil droplets dispersed in the aqueous phase. This ensures good wetting of the metal surface by the oil as well as good stability for the emulsion in cases in which the MWF are recirculated for reuse. Here, surfactants with moderate to long hydrophobes with good oil/water interfacial tension reductions are used. Furthermore, the surfactants should have the ability to release the oil phase by destabilizing the emulsion once the emulsion comes in contact with the cutting surfaces during the metalworking operations.

Salts of phosphate ester and fatty acid-based anionic surfactants with C_{12} to C_{18} alkyl chains are used extensively in metalworking fluids to provide lubricity. The ability of these surfactants to complex strongly with metals ensures a tightly packed monomolecular film, with the hydrophobes oriented away from the metal surface. Moles of ethoxylation for these surfactants range from 0 to 6 mol, as any further increase would reduce both their lubricity as well as their solubility in petroleum-based MWF. In addition, a major benefit, particularly of the phosphate ester surfactants, is their ability to interact chemically with the metals through their phosphorus atom to form a low-melting eutectic metal phosphide film on the metal surface. Phosphide films, being soft and malleable, provide an interface that minimizes welding and friction between metal surfaces, thereby reducing excessive wear and pitting on both the metal and the cutting tools.

The majority of surfactants used as emulsifiers in the formulation of soluble and semi-synthetic MWF are nonionic or anionic. It is usual to find that the majority of anionic surfactants that are used to provide lubricity are also optimized with respect to their hydrophobicity and hydrophilicity in order to function as good emulsifiers as well as to reduce the substrate/oil interfacial tensions. In this respect, phosphate esters with C_{12} to C_{18} alkyl chains that have olefinic or aryl groups are commonly used. Moles of ethoxylation and the alkyl chain length for these surfactants are optimized to provide good solubility in petroleum solvents, and in the case of soluble oils, lubricity and emulsifier properties. Nonionic surfactants with branched C_8 to C_9 alkylated phenol and branched C_{10} to C_{13} with moles of ethoxylation ranging from 5 to 10 are frequently used as emulsifiers in the formulation of both soluble and semi-synthetic MWF. The branched chains for these surfactants effectively reduce the oil/water interfacial tension required for good emulsification of the MWF concentrates as well as good emulsification of the formulations on dilution with water. Further, these nonionic surfactants, on reaching their cloud point temperatures during metalworking operations, destabilize or break up the emulsions, effectively releasing the oil phase onto the metal surface, which ensures good lubricity for the MWF. Surfactants used commonly in metalworking fluids are listed in Tables 7.12 and 7.13.

Plastics

Surfactants play a major role for both fabricators and end users in providing some desirable properties that enhance the characteristics and utilities of plastics. They are used in plastic manufacture as antistatic, mold release, plasticizing, antiblocking, and defogging agents.

Antistatic Agents

During the processing of plastics, the synthetic resins are exposed to severe friction, which produces the buildup of static charges onto plastics that are responsible for certain problems: (i) attracting dirt particles that impair appearance, (ii) giving unpleasant shocks to people, and (iii) causing adhesion of resin and plastics to the molds and machinery parts.

TABLE 7.12
Surfactants Used in Metalworking Fluids

Surfactants used as lubricants	Metalworking fluids
$[C_{12}H_{25}(OC_2H_4)_{0-4}O]_{1,2}P(O)(O^-X^+)_{1,2}$ [a]	Straight and soluble oils
$[(br)C_{13}H_{27}(OC_2H_4)_{0-4}O]_{1,2}P(O)(O^-X^+)_{1,2}$	
$C_{18}H_{35}(OC_2H_4)_{0-5}O]_{1,2}P(O)(O^-X^+)_{1,2}$	
$[p\text{-}t\text{-}C_9H_{19}C_6H_4(OC_2H_4)_{2-4}O]_{1,2}P(O)(O^-X^+)_{1,2}$	
$[(C_9H_{19})_2C_6H_3(OC_2H_4)_{6-10}O]_{1,2}P(O)(O^-X^+)_{1,2}$	
$C_{14-16}H_{29-33}COO^-X^+$	
$C_{16-18}H_{33-37}COO^-X^+$	
$C_{18}H_{35}COO^-X^+$	
$C_{18}H_{37}COO^-X^+$	
$[(br)C_{13}H_{27}(OC_2H_4)_{4-6}O]_{1,2}P(O)(O^-X^+)_{1,2}$	Semi-synthetic
$C_{18}H_{35}(OC_2H_4)_{4-8}O]_{1,2}P(O)(O^-X^+)_{1,2}$	
$[p\text{-}t\text{-}C_9H_{19}C_6H_4(OC_2H_4)_{5-8}\ O]_{1,2}P(O)(O^-X^+)_{1,2}$	
$[[C_6H_5CH(CH_3)_2]_3C_6H_2O(C_2H_4O)_{5-15}O]_{1,2}P(O)(O^-X^+)_{1,2}$	
$[C_4H_9(OC_2H_4)_{4-6}O]_{1,2}P(O)(O^-X^+)_{1,2}$	Synthetic
$[C_6H_5(OC_2H_4)_{4-8}O]_{1,2}P(O)(O^-X^+)_{1,2}$	
$[CH_3C_6H_4(OC_2H_4)_{4-8}O]_{1,2}P(O)(O^-X^+)_{1,2}$	
$[(br)C_8H_{17}(OC_2H_4)_{4-6}O]_{1,2}P(O)(O^-X^+)_{1,2}$	

[a]Abbreviation: X = K, Na, $NH_2CH_2CH_2OH$, $NH(CH_2CH_2OH)_2$, $N(CH_2CH_2OH)_3$.

In order to be effective in providing good antistatic properties, the surfactant selected should retard or prevent the buildup of static electricity or promote its rapid discharge. In order to do this, the surfactants used should have the ability to migrate to the surface of the plastic with the orientation of their polar hydrophilic groups exposed to the atmosphere to form an efficient conductive ionic film on the surface of the plastic. Further, the surfactant used should have good compatibility with the plastic and good heat stability to withstand temperatures >500°F, and should resist decomposition, volatilization, and oxidation. In addition to providing antistatic properties, the selection of surfactants is also very often based on their ability to provide plasticizing, lubricity, and mold release properties.

Due to their ability to provide almost all of these desired properties, phosphate ester anionic surfactants are the most popularly used antistatic agents in the manufacture of plastics. They withstand high temperatures and show good compatibility toward most commercially used plastics, such as polyethylene, polypropylene, polystyrene, ABS, and PVC. Nonionic surfactants, due to their low thermal stability, are rarely used in

TABLE 7.13
Emulsifiers Used in Metalworking Fluids

Surfactants used as emulsifiers	Metalworking fluids
$C_{18}H_{35}(OC_2H_4)_{3-6}O]_{1,2}P(O)(O^-X^+)_{1,2}$ [a]	Soluble oils
$[p\text{-}t\text{-}C_9H_{19}C_6H_4(OC_2H_4)_{4-6}O]_{1,2}P(O)(O^-X^+)_{1,2}$	
$[(C_9H_{19})_2C_6H_3(OC_2H_4)_{8-10}O]_{1,2}P(O)(O^-X^+)_{1,2}$	
Castor oil $(OC_2H_4)_{15-40}OH$	
$[p\text{-}t\text{-}C_9H_{19}C_6H_4(OC_2H_4)_{6-10}O]_{1,2}P(O)(O^-X^+)_{1,2}$	Semi-synthetics
$[(C_9H_{19})_2C_6H_3(OC_2H_4)_{10-15}O]_{1,2}P(O)(O^-X^+)_{1,2}$	
$C_{10-12}H_{21-25}O(C_2H_4O)_{2-6}C(O)O^-X^+$	
$C_{12-14}H_{25-29}(OC_2H_4)_{0-10}OC(O)O^-X^+$	
$C_{16-18}H_{33-37}O(C_2H_4O)_{0-15}C(O)O^-X^+$	
$p\text{-}t\text{-}C_9H_{19}C_6H_4(OC_2H_4)_{4-10}OH$	
$p\text{-}t\text{-}C_8H_{19}C_6H_4(OC_2H_4)_{4-8}OH$	
(br)$C_{13}H_{27}(OC_2H_4O)_{6-10}OH$	
$C_9H_{19}C(CH_3)_2S(C_2H_4O)_{6-8}H$	
$C_{12}H_{25}C(O)N(CH_2CH_2OH)_2$	
$C_{16-18}H_{32-35}C(O)N(CH_2CH_2OH)_2$	

[a]Abbreviation: X = K, Na, $NH_2CH_2CH_2OH$, $NH(CH_2CH_2OH)_2$, $N(CH_2CH_2OH)_3$.

high-temperature plastic processing. Anionic surfactants such as sulfates and sulfonates show poor compatibility and are not used. Cationic surfactants are generally used only as external antistatic agents due to their poor compatibility toward plastics and poor thermal stability.

Among phosphate ester surfactants, those with C_{12} to C_{14} alkyl chains, preferably branched and aryl-based hydrophobes, are used most commonly. Although surfactants with longer alkyl chains are much more compatible with plastics, they provide much slower migration to the plastic/air interface and, as such, are less effective in imparting antistatic properties. Branched alkyl hydrophobes provide good controlled release of the surfactant to surfaces as well as good plasticizing and mold release properties. In order to provide good surface migration and to form a conductive ionic film, phosphate ester surfactants used contain polyoxyethylene chains with 4 to 8 oxyethylene units. With more than 9 to 10 oxyethylene groups, these surfactants become incompatible with the plastics, which causes blooming that results in excessive condensation of moisture on the surface. To increase compatibility, phosphate ester surfactants with a diester/monoester ratio >50% are generally preferred.

Surfactants most commonly used are listed in Table 7.14. In the case of polystyrene, the ideal phosphate ester surfactants are those that are based

TABLE 7.14
Surfactants Used as Antistats in Plastic Manufacture

$[C_{12}H_{25}(OC_2H_4)_{2-4}O]_{1,2}P(O)(OH)_{1,2}$
$[(br)C_{13}H_{27}(OC_2H_4)_{4-6}O]_{1,2}P(O)(OH)_{1,2}$
$[C_9H_{19}C_6H_4(OC_2H_4)_{4-6}O]_{1,2}P(O)(OH)_{1,2}$
$[(C_9H_{19})_2C_6H_3(OC_2H_4O)_{6-10}]P_{1,2}(O)(OH)_{1,2}$
$[C_{18}H_{35}(OC_2H_4O)_{4-6}]P_{1,2}(O)(OH)_{1,2}$
$R–N[(CH_2CH_2OH)_x]_2$, $R = C_{12}–C_{18}$, $x = 2–4$
$CH(OH)(CH_2OC(O)CR)_2$, $R = C_{11}–C_{17}$
$R–N^+(CH_3)_3C1^-$, $R = C_{16}–C_{18}$
$R–N^+(CH_3)_2CH_2COO^-$, $R = C_{16–18}$ or $C_{15–17}$ $C(O)NH(CH_2)_3$
$RC–(O)NH(CH_2)_2NH^+(CH_2CH_2OH)CH_2COO^-$, $R = C_{11}–C_{17}$

on nonyl or octylphenol ethoxylates. This preference may be attributed to their good compatibility due to possible interaction between the aryl groups of the surfactants and the polymers.

Slip and Mold Release Agents

Processing of resins with high tackiness, such as high-density polyethylene, polypropylene, or polyolefin, is plagued by processing problems. The tackiness causes a great increase in the friction between the processing equipment and the polymer. Further, if tackiness is not eliminated it is carried onto the finished product, especially in the processing of polyolefin films in which tackiness causes roller-wrapped films to adhere to each other.

Here too, as in the case of antistatic agents, the surfactants should show good compatibility and solubility in the resin at high temperatures as well as the ability to migrate to the surface in order to form a polar or a hydrophilic film on the surface of the plastic. This would eliminate or minimize surface tackiness and prevent surfaces from sticking together. Surface modification allows these surfactants to function as internal mold release agents as well. They speed the processing of plastic molding and extruding operations as well as provide molded products with smooth surfaces and superior gloss. In addition, the selection of surfactants as mold release slip agents is also based on their having good lubricity and color stability at high processing temperatures.

Alkanolamides and phosphate ester surfactants with saturated or unsaturated long-chain C_{18} to C_{22} alkyl chains are used in these applications. Amides with unsaturated alkyl chains (oleic, linoleic, and euracyl) are used where color and temperature stability are not required. Saturated alkanolamides are used when processing plastics with very high melting points and where tackiness is a major problem. They prevent plastic containers and rolls of plastic film from adhering to each other when stacked. Amides are preferred over other nonionic and anionic surfactants due to their excellent lubricity, compatibility as well as their high temperature stability and low volatility. The nitrogen atoms in these amides, being electron rich, provide good repulsion between the molds and the film surfaces to provide good slip and mold release properties. Commonly used surfactants for these applications are listed in Table 7.15.

Defogging Agents

The hydrophobicity, and therefore the poor aqueous wettability, of plastics often causes moisture to condense as water droplets on the surfaces of plastic films and panels. Condensation of moisture as water droplets

TABLE 7.15
Surfactants Used in Plastic Processing for Slip and Mold Release and Antiblock Properties

Surfactant	Applications
$C_{15-17}H_{31-35}C(O)NH_2$, $C_{15-17}H_{31-35}C(O)NHCH_2CH_2NHCH_2CH_2OH$	Slip agent for polypropylene and polyethylene film
$C_{13-17}H_{31-35}C(O)NH_2$	Antiblock and mold release for high temperature injection molding of polyvinylidene chloride and PVC resins
$C_{21}H_{43}C(O)NHCH_2CH_2OH$, $C_{21}H_{43}C(O)NHCH_2CH_2OH$	Exceptional lubricity, antislip mold release for PVC and polypropylene films used in high melting plastics
$[C_{15-17}H_{31-35}(OC_2H_4)_2O]_{1,2}P(O)(OH)_2$	Mold release agent and plasticizer for all types of plastics
$[C_{15-17}H_{31-35}C(O)NHCH_2CH_2NHCH_2]_2$	Mold release and lubricant extrusion aid for injection molding of PVC
$(C_8H_{17}O)_3P(O)$	Plasticizer and tack reduction in vinyls (PVC) at high temperatures

obscures visibility due to the poor penetration of light and, in the case of plastics used for food packaging, this causes poor consumer appeal for the products. In situations in which rigid plastic panels are used for enclosures in greenhouses, water condensate reduces the flow of solar light onto the plants. In order to eliminate or minimize the condensation of water as droplets, plastics are very often treated during their manufacture with defogging agents that are blended in during processing.

The most important interfacial property in preventing fogging on plastics is wetting; consequently, the most desired surfactant property is reduction of the substrate/aqueous interfacial tension. Since the substrates are mostly hydrophobic in nature, the surfactants must have the ability to migrate to the surface of the plastic and be oriented in such fashion as to lower the substrate/aqueous solution interfacial tension. Also, in selecting surfactants it is essential that they demonstrate slow release to the surface and firm anchoring to the plastic surface in order to prevent bleeding or being washed away. Good heat and color stability, in addition to maintaining good transparency of the plastic, are essential properties in selecting surfactants as defogging agents.

Among the surfactants most commonly used are polyoxyethylenated or polyhydroxylated surfactants with alkyl chain lengths having 9 to 12 carbon atoms. Surfactants having aromatic bonds in their hydrophobes are preferred, as they appear to provide good substantivity and slow release, and therefore good defogging properties, particularly toward PVC and polystyrene-based plastics. The moles of ethoxylation for these nonionics range from 0 to 4. With higher ethoxylation, the compatibility of surfactants toward plastics decreases, which causes bleeding onto the surface of the plastic. The most commonly used surfactants are listed in Table 7.16.

Recovery of Surfactants for Reuse in Industrial Cleaning Operations

Restrictions in the use of chlorinated solvents for industrial metal cleaning have significantly increased the demand for aqueous cleaners. With the growth of aqueous cleaners, the disposal of wastewater generated in the industrial cleaning operations has become a major problem, both economically and environmentally. Although there are a number of technologies available to treat industrial waste, the most effective and fastest growing is the use of ultrafiltration membrane units. During ultrafiltration, the contaminated wash water, mostly in the form of O/W emulsions, is pumped through 0.05 to 0.1 μm pore size inorganic or organic filters that allow

TABLE 7.16
Surfactants Used as Defogging Agents in Plastics

(br)$C_9H_{19}C_6H_4\ O(C_2H_4O)_{3-6}H$
(br)$C_{11-13}H_{23-27}CH(CH_3)O(C_2H_4O)_{3-6}H$
(br)$C_{13}H_{27}O(C_2H_4O)_{4-6}H$
$CH(OH)[CH_2O(O)C\ C_{11}H_{23}]_2$
$C_{17}H_{35}C(O)O(CH_2CH_2O)_{1-4}H$

only the passage of an aqueous phase. Oil contaminants are removed by the filtration and are concentrated for disposal.

In order for the treatment of industrial wastewater by the ultrafiltration process to be economically viable, it is essential that most of the surfactants in the wastewater be recovered in the filtrate (permeate) along with the aqueous phase so that the recycled aqueous phase maintains its cleaning performance. Any loss of surfactants in the permeate must be supplemented by adding fresh surfactants to the cleaner bath, as the depletion of surfactants will continue each time the cleaner bath is sent through the ultrafiltration membrane unit.

The majority of conventional products used in industrial cleaners are formulated with nonionic surfactants, due to their ability to provide good cleaning power as well as their ability to emulsify heavy oils and greases. However, studies and field experience have shown that the majority of these surfactants are removed with the oil phase upon ultrafiltration. Table 7.17 shows a typical nonionic-based cleaner formulation used in aqueous based cleaning applications (21).

Table 7.18 shows the percentage of nonionic in this cleaner passing though an ultrafiltration membrane and present in the permeate. After only 5 min (one pass), almost 90% of the cleaner has been "stripped out" by the ultrafiltration process. Therefore, the economics of incorporating

TABLE 7.17
Nonionic-Based Aqueous Cleaner

Component	Weight Percentage
Water	84.5
Potassium phosphate	4.0
Sodium metasilicate	1.5
Igepal CO-630 (nonoxynol-9)	4.0
Sodium xylene sulfonate (40% solution)	6.0

TABLE 7.18
Percentage of Nonionic Surfactant in the Cleaner Passing Through the 0.1 μm Pore Size Ultrafiltration Membrane, Initial Surfactant Concentration (Ci) = 3.7%; pH = 12

Time (min) the cleaner solution is passed through the filter	Nonionic surfactant concentration in permeate (Cp) (wt%)	(Cp) (wt%)
5	0.316	10.5
30	0.309	10.3
60	0.302	10.1
120	0.288	9.6
240	0.251	8.4

this type of cleaner into an ultrafiltration process is not efficient both for performance and economic reasons.

Consequently, the surfactants used in ultrafiltration processes must, in addition to providing detergency, have good permeability along with the aqueous phase when subjected to ultrafiltration. In order for the surfactants used in the industrial cleaner formulation to be recovered and reused by ultrafiltration, the surfactants should show (22):

- the ability to form metastable O/W emulsions that separate easily when subjected to interfacial turbulence due to capillary pressure.
- good solubility at all temperatures, with no "cloud point" phase separation behavior.
- good solubility in the presence of hard water and of the heavy metal ions found in metal cleaning baths.
- minimal adsorption onto the ultrafiltration filter surfaces. In general, the majority of ultrafiltration membranes are hydrophilic, and any adsorption of surfactants makes the surfaces hydrophobic, preventing wetting of the filter surface and good permeability by the aqueous solution.

Due to their ability to provide almost all of the above desired properties, the use of zwitterionic surfactants in industrial cleaner formulations is growing rapidly. These surfactants, especially those that carry charges at all pHs, show exceptionally good solubility in the aqueous phase and very limited solubility in the oil phase at all temperatures. Furthermore, they are extremely compatible with heavy metal ions and hard water, thereby maintaining good solubility under these conditions. Studies have

TABLE 7.19
Percentage of Mixed C_4 and C_8 Alkylether Hydroxy Propyl Sultaine Surfactant RO CH_2 CH (OH) CH_2 N^+ $(CH_3)_2$ CH_2 CH (OH) CH_2 SO_3^- Passing Through 0.1 μm Pore Size Membrane—Initial Surfactant Concentration (C*i*) = 0.25%; pH = 12

Time (min) the cleaner solution is passed through the filter	Surfactant concentration in permeate (Cp) (wt%)	Cp/Ci (%) (wt%)
5	0.175	70.0
30	0.197	78.8
60	0.164	65.6
120	0.175	70.0
240	0.135	54.0

shown that formulations containing them maintain 100% conductivity (with respect to water) through the filters and show minimal adsorption onto the filter surfaces, thereby avoiding any plugging of filter pores. Among zwitterionic surfactants, those with C8 to C12 alkyl chains provide optimum detergency and recovery. Sultaines are preferred over betaines due to their exceptionally good compatibility with the heavy metal ions found in metal cleaning baths. They also form metastable O/W emulsions that break up easily at the filter surface, with the surfactant partitioning mostly into the aqueous phase. Their zwitterionic nature at all pHs in the presence of heavy metal ions prevents them from adsorbing onto and plugging the filter surfaces. Tables 7.19 and 7.20 illustrate the percentage of these amphoteric/zwitterionic surfactants passing through

TABLE 7.20
Percentage of C_{8-10} Alkylaminopropionate, $RCOCH_2CH_2N^+H(CH_2CH_2COO^-)CH_2CH_2COO^-)NA^+$ Surfactant Passing Through 0.1 μm Pore Membrane. Initial Surfactant Concentration (C*i*) = 5.7% (active)

Min	Concentrate of permeate (CP) (wt% active)	Cp/Ci (%) (wt % active)
5	5.12	90.0
30	5.50	97.1
60	5.68	98.9
120	5.10	91.0
240	5.08	90.2

TABLE 7.21
Surfactants Used in Ultrafiltrable Cleaners

$R\ CH\ (C_2H_5)\ CH_2N^+\ (CH_3)_2\ CH_2\ CH\ (OH)\ CH_2SO_3^-$
$R\ CH\ (C_2H_5)\ CH_2O\ CH\ (OH)\ CH_2N^+\ (CH_3)_2\ CH_2\ CH\ (OH)\ CH_2\ SO_3^-$
$R\ CH\ (OH)\ CH_2N^+\ (CH_3)_2\ CH_2\ CH\ (OH)\ CH_2\ SO_3^-$
$R\ C\ (O)\ NH\ (CH_2)_3\ N^+\ (CH_3)_2\ CH_2\ CH\ (OH)\ CH_2\ SO_3^-$
$R\ O\ CH_2\ CH\ (OH)\ CH_2N^+\ (CH_3)_2\ CH_2\ CH\ (OH)\ CH_2\ SO_3^-$
$R_1\ N^+\ (H)\ (CH_2\ COO^-)\ (CH_2\ COO_2^-\ Na^+)$
$R_1\ N^+\ (H)\ (CH_2\ CH_2\ COO^-)\ CH_2\ CH_2\ COO\text{–}\ Na^+$
$R_1\ N^+\ (CH_3)_2\ O^-$
$R = C_6$ to C_{12}
$R_1 = C_5\ CO\ CH_2\ CH_2$ to $C_{11}\ CO\ CH_2\ CH_2$ or C_6 to C_{12}

the ultrafiltration membrane and collected in the permeate (21, 22). The most commonly used surfactants are listed in Table 7.21.

References

1. Glass Fiber Dispersions, Sheets Plastic Impregnated Sheets and Method of Forming, Hungerford, G.P., US Patent 2,906,660 (1959).
2. Fibrous Glass Product and Method of Manufacture, Slayter, G., Morgan and Morrison, A.R., US Patent # 3,050,427 (1962).
3. Process for Manufacturing Endless Fiber Webs from Inorganic Fiber Suspensions, Shuller, W., Werner, H., US Patent # 3,766,003 (1973).
4. Method and Means for Strand Filament Dispersal, Pitt, R.E., US Patent # 3,760,458 (1973).
5. Filamentization Process For Glass Fibers. Stelego, J.P., US Patent # 3,634,054 (1972).
6. Glass Fiber Dispersants for Making Uniform Glass Fiber Mats by the Wet-Laid Process. Chakrabarti, P.M., US Patent # 4,179,331 (1979).
7. Jakush, E.A., Dispersants in Wet Laid Glass Mat, *Tappi Proceedings, 1991 Nonwoven Conference*, p. 163.
8. Use of Propoxylated Fatty Amine Ethoxylates as Glass Fiber Dispersants for the Manufacture of Uniform Glass Fiber Mats, Razak, S., and P. Eckler, US Patent # 5,409,574 (1995).
9. Amphoteric Surfactants as Glass Fiber Dispersants for the Manufacture of Uniform Glass Fiber Mats. Dahanayake, M., Razak, S., Rhone Poulenc Inc., US Patent # 5,407,536 (1995).
10. Razak, S., and M., Dahanayake, Novel Fatty Amine Alkoxylate Surfactants as Glass Fiber Dispersants for the Manufacture of Uniform Glass Fiber Mats, Rhone Poulenc Inc., *Tappi Proceedings, 1995 Nonwoven Conference*, Gainsville, Fla., p. 163.

11. Hughes, T., T. Jone, and G. Tustin, Viscoelastic Surfactant-Based Gelling Composition for Wellbore Service Fluids, Patent UK GB 2,332,223 (1999).
12. Brown, J.E., R.J. Card, and E. Nelson, Methods and Compositions for Testing Subterranean Formulations. U.S. Patent 5,964,295.
13. Norman, E.C., C. Springs, and A.C. Regina, Coatesville, Alcohol Resistant Aqueous Film Forming Firefighting Foam, US Patent # 4,999,119 (1991).
14. Mulligan, D.J., Bentham, Lancaster, Fire-Fighting Compositions, US Patent # 4,424,133 (1984).
15. Alm, R.R., Lake Elmo, and R.M. Stern, Woodbury, Aqueous Film-Forming Foamable Solution Useful as Fire Extinguishing Concentrate, US Patent # 5,085,786 (1992).
16. Falk, R.A., Perfluoralkyl Anion/Perfluoroalkyl Cation Ion Pair Complexes, US Patent # 4,420,434 (1983).
17. Falk, R.A., Perfluoroalkyl Anion/Perfluoroalkyl Cation Ion Pair Complexes, US Patent # 4,472,286 (1984).
18. Anderson, R.D., Process for Rapidly Dissolving Water-Soluble Polymers, US Patent # 3,624,019 (1971).
19. Anderson, R.D., O. Frisque, Rapidly Dissolving Water-Soluble Polymers, US Patent # 3,734,873.
20. Dahanayake, M., E. Larson, Rhone Poulenc Inc., Surfactants for Self Inverting Polyacrylamides, US Patent # 5,925,714 (1999).
21. Dahanayake, M., and B. Yang, Recovery and Reuse of Surfactants from Aqueous Solutions. U.S. Patent 5,654,480 (1997).
22. Ventura, M., and M. Dahanayake, Filterable Surfactant Class Resolves Separation Anxiety, *Precis. Clean. Mag.* (July, 1998).

Major Surfactant Suppliers

AKZO NOBEL CHEMICALS INC

Chemical name	Trade name
Anionic	
Alpha olefin sulfonates	Elfan
Alkylbenzene sulfonates	Berol, Elfan
Alcohol sulfates	Elfan
Alkyl ether sulfates	Elfan
Alkyl sulfosuccinates	Elfanol
Alkylisethionates	Elfan
Nonionic	
Fatty alcohol ethoxylates	Berol
Fatty acid ethoxylates	Berol
Alkylphenol ethoxylates	Berol
Fatty acid esters of glycerin	Homotex
Fatty acid esters of polyethanol glycol	Homotex
Alkyl amine ethoxylates	Ethomeen
Fatty Acid diethanol amides	Ethomid
Cationic	
Alkyl quaternary ammonium salts	Arquad
Dialkyl quaternary ammonium salts	Arquad
Polyoxyethylenated quaternary ammonium salts	Ethoquad
Diquaternary ammonium dichloride	Duoquad
N-alkyltrimethylene diamine diacetate	Duomac
Amphoteric and Amine Oxides	
Alkyl amphoglycinates	Ampholak
Alkyl iminoglycinate	Ampholak
Alkyl amphopolycarboxy glycinates	Ampholak
Alkyliminodipropionate	Ampholak

BASF CORPORATION, SPECIALTY CHEMICALS

Chemical name	Trade name
Anionic	
Alkyl benzene sulfonic acid	Mazon
Phosphate ester ethoxylates	Klearfac, Maphos
Alkylphenol ether phosphate esters	Emulan, Baspon
Cocoisethionate	Jordapon
Nonionic	
Fatty acid ethoxylates	Mapeg, Cremophor
Alkyl alcohol ethoxylates	Iconol
Alkylphenol ethoxylates	Iconol
Castor oil ethoxylates	Cremophor

BASF CORPORATION, SPECIALTY CHEMICALS (*Continued*)

Chemical name	Trade name
Fatty esters of sorbitan	S-Maz
Fatty esters of glycerol	Cremophor, Mazol
Fatty acid esters of ethoxylated glycerols	Cremophor
Fatty acid esters of ethoxylated sorbitan	T-Maz
Fatty amine ethoxylates	Icomeen
Block co-polymers of ethylene oxide and propylene oxide	Pluronic, Pluronic R
Alcohol alkoxylates	Plurafac
Ethylenediamine alkoxylates	Tetronic, Tetronic R
Trimethylopropane alkoxides	Pluradot

COGNIS CORPORATION

Chemical name	Trade name
Nonionic	
Alcohol ethoxylates	Dehydol, Trycol
Alkylphenol ethoxylates	Trycol
Fatty alcohol ethoxylates	Emulgin, Trycol
Fatty acid ethoxylates	Dehymuls, Emerest
Fatty acid alkoxylates	Dehypon
Fatty acid esters of glycerol	Emerest
Fatty acid esters of ethoxylated glycerol and glycol	Emerest
Fatty acid esters of ethoxylated sorbitol	Dehymuls
Fatty amine ethoxylates	Trymeen
Alkyl diethanolamides	Standamide, Emid
Alkanol superamides	Standamid
Alkylpolyglycosides	Glucopon
Anionic	
Alkyl sulfates	Standapol
Alkyl ether sulfates	Standapol
Alkylphenol ether sulfates	Standapol
Alkyl naphthalene sulfonates	Sellogen
Alkyl sulfosuccinates	Standapol
Betaines, Amphoterics and Amine Oxides	
Alkyl amphoglycinates	Velvetex
Alkyl amidopropyl betaines	Dehyton, Velvetex
Alkyl amidopropyl amine oxide	Standamox
Alkyl amidopropyl sulfobetaine	Dehyton, Velvetex
Alkyl imino dipropionate	Deriphat
Alkyl amine oxide	Standamox
Cationic	
Alkyl quaternary ammonium salts	Dehyquart

CONDEA VISTA

Chemical name	Trade name
Anionic	
Alkyaryl sulfonates	Marlon, Maropon, Sermul
Petroleum sulfonates	Marlon

CONDEA VISTA (*Continued*)

Chemical name	Trade name
Fatty alcohol sulfates	Serdet
Fatty alcohol ethersulfates	Marlinat, Sermul
Alkylaryl ethersulfates	Sermul
Alkyl ether phosphate esters	Servoxyl
Alkylaryl phosphate esters	Marlon
Alcohol ether carboxylates	Marlinat, Marlowet
Sulfonated naphthalene formaldehyde condensate	Sermul
(Di) alkyl sulfosuccinates	Sermul, Serwet
Nonionic	
Fatty alcohol ethoxylates	Alfonic, Marlipal, Marlowet, Marlox
Fatty acid ethoxylates	Marlowet, Servirox
Fatty alcohol alkoxylates	Marlox
Fatty acid alkoxylates	Marlox
Alkylphenol ethoxylates	Marlophen
Fatty amine ethoxylates	Marlazin, Marlowet, Serdox, Servon
Alkyl diethanol amides	Serdolamide
Alkyl imidazoline	Serramine, Marlowet, Sermul
Cationic	
Quaternary amino polyglycol ether	Seramine
Polyquaternary ammonium salts	Seramine
Fatty acid esters of glycerol and glycol	Alkamuls

CROMPTON CORPORATION

Chemical name	Trade name
Anionic	
Alpha olefin sulfonates	Witconate
Linear alkylaryl sulfonates	Witconate, Hyb
Branched alkylbenzene sulfonates	Witconate
Petroleum sulfonates	Witconate, Petronate
Alkyl naphthalene sulfonates	Petronate
Alkylphenol ethersulfates	Witconate
Alkyl sulfates	Witcolate
Alkyl ether sulfates	Witcolate
Alkyl sulfosuccinates	Emcol
Alkyl ether sulfosuccinates	Varsulf
Alkyl ether phosphate esters	Emphos, Desophos
Alkylphenol ether phosphate esters	Emphos
Fatty acids	Industrene
Fatty acid ether carboxylates	Emcol
Nonlonic	
Alcohol ethoxylates	Varonic
Alkylphenol ethoxylates	Desonic, Armul
Fatty acid ethoxylates	Desonic
Alkyl amine ethoxylates	Varonic, De Someen, Witcameen
Alkyl monoethanol amides	Varamide, Witcamide

CROMPTON CORPORATION (*Continued*)

Chemical name	Trade name
Alkyl diethanol amides	Varamide, Witcamide
Fatty acid esters of glycerol and glycol	Kemester
Fatty acid esters of sorbitan	Armul
Fatty acid esters of ethoxylated sorbitan	Desotan
Silicone surfactants	Silwet
Amphoterics, Betaines and Amine Oxides	
Alkyl betaines	Rewoteric, Emcol
Alkyl amidopropyl betaine	Rewoteric
Alkyl aminopropyl sulfobetaines	Rewoteric
Alkyl amphoacetate	Rewoterlc
Alkyl amphopropionates	Rewoterlc
Alkyl aminoethyl imidazoline	Rewoteric
Alkyl amphohydroxysultaines	Rewoterlc
Alkyl amine oxides	Varox
Cationic	
Polypropoxy quaternary ammonium salts	Emcol
Lapyrium salts	Emcol
Dialkyl quaternary ammonium salts	Variquat
Alkyl benzyl ammonium salts	Variquat

DOW UNION CARBIDE

Chemical name	Trade name
Nonionic	
Alkyphenol ethoxylates	Tergitol, Triton CF, Triton X, Triton NP
Secondary (br) alcohol alkoxylates	Tergitol Min-Foam, Tergitol TMN, Tergitol S
Polyalkylene glycol ethers	Tergitol XD, Tergitol XH
Block copolymers of EO/PO	Tetronic
Alkylpolyglucosides	Triton BG
Cleavable ethoxylates	Triton SP
Anionic	
Alkylaryl ether sulfates	Triton W, Triton XN
Alkyl ether sulfates	Triton QS
Alkyl diphenyloxide disulfonate	Dowfax
Alkyl sulfosuccinates	Triton GR
Alkyl phosphate ester	Triton H & QS

HUNTSMAN CORPORATION

Chemical name	Trade name
Anionic	
Linear alkylbenzene sulfonates	Surfonic SNS
Branched alkylbenzene sulfonates	Surfonic SNS
Alkyl sulfosuccinates	Surfonic DOS
Alcohol ether sulfates	Surfonic SB
Alkyl ether phosphates esters	Surfonic PE

HUNTSMAN CORPORATION (*Continued*)

Chemical name	Trade name
Nonionic	
Alkylphenol ethoxylates	Surfonic DDP
Alcohol ethoxylates	Surfonic DA
Alcohol alkoxylates (EO/PO)	Surfonic JL & LF
Fatty acid alkoxylates (EO/PO)	Surfonic TX-RE
Fatty acid esters of glycerol	Surfonic E
Block co-polymers of ethylene oxide-propylene oxide	Surfonic POA
Alkyl monoethanol amides (1:1)	Surfonamide LMEA
Alkyl diethanol amides	Surfonamide OD
Alkanol superamides	Surfano amide
Alkyl amine ethoxylates	Surfonic T

RHODIA

Chemical name	Trade name
Anionic	
Fatty alcohol sulfates	Rhodapon
Fatty alcohol ether sulfates	Rhodapex
Tristyrylphenol ether sulfates	Rhodapex
Alpha olefin sulfonates	Rhodacal
Alkylbenzene sulfonates	Rhodacal
Dialkylphenyl ether disulfonate	Rhodacal
Alkyl naphthalene sulfonate	Supragil
Sulfonated naphthalene formaldehyde condensate	Supragil
(Di) Alkyl sulfosuccinates	Gerapon
(Mono) Alkyl sulfosuccinates	Gerapon
Alkyl sulfosuccinamates	Gerapon
Fatty acid n-methyl taurates	Gerapon
Fatty acid isethionates	Gerapon
Mono alkylether phosphate esters (MAPS)	Rhodafac
Di alkyl ether phosphate esters	Rhodafac
Alkylphenol ether phosphate esters	Rhodafac RM
Di alkylphenol ether phosphate esters	Rhodafac
Tristyrylphenol ether phosphate esters	Rhodafac
Alkyl ether carboxylates	Miranate
Nonionic	
Linear alcohol ethoxylates	Rhodasurf
Branched alcohol ethoxylates	Rhodasurf
Alkylphenol ethoxylates	Igepal
Dialkylphenol ethoxylates	Igepal
Tristyrylphenol ethoxylates	Soprophor
Nopol alkoxylates	Rhodoclean
Alkyl thioethoxylates	Alcodet
Alcohol alkoxylates (EO/PO)	Antarox
Block copolymers of (EO/PO)	Antarox
Fatty acid ethoxylates	Rhodasurf
Fatty acid esters of sorbitan	Alkamuls
Fatty acid esters of ethoxylated sorbitan	Alkamuls
Fatty acid esters of glycerol and glycol	Alkamuls

RHODIA (*Continued*)

Chemical name	Trade name
Fatty acid esters of polyethylene glycol	Alkamuls
Alkyl superamides	Antarox AE33
Alkyl diethanol amides (1:1)	Alkanolamide
Alkyl diethanol amides (2:1)	Alkanolamide
Alkyl monoethanol amides	Alkanolamide
Alkyl isopropanolamides	Alkanolamide
Fatty amine ethoxylates	Rhodameen
Amphoteric and zwitterionic	
Alkyl amphoacetates	Miranol
Alkyl amphodiacetates	Miranol
Alkyl amphodipropionate	Miranol
Alkyl imino dipropionate	Mirataine
Alkyl amidopropyl betaine	Mirataine
Alkyl amido hydroxypropyl sultaine	Mirataine
Alkyl ether hydroxypropyl sultaine	Mirataine
Alkyl amido propyl amine oxide	Rhodamox
Alkyl amine oxide	Rhodamox
Cationic	
Alkyl hydroxyethyl imidazoline	Miramine C
Alkyl quaternary ammonium salts	Alkaquat, Rhodaquat
Alkyl benzyl ammonium salts	Alkaquat, Rhodaquat
Alkyl pyridinium ammonium salts	Alkaquat, Rhodaquat
Dialkyl quaternary ammonium salts	Alkaquat, Rhodaquat
Ditallow imidazolium salts	Alkaquat, Rhodaquat

SHELL CHEMICAL COMPANY

Chemical name	Trade name
Nonionic	
Linear alcohol ethoxylates	Neodol
Fatty acid alkoxylates	Nontell

STEPAN COMPANY

Chemical name	Trade name
Anionic	
Alkyl and alkylarylbenzene sulfonates	Ninate
Alkyl benzenesulfonates	Biosoft, Polystep A
Alkyl sulfates	Stephanol, Polystep B
Alkyl ether sulfates	Steol
Alkyl sulfosuccinates	Anionyx
Alkyl sarcosinates	Bioterge
Alpha olefin sulfonates	Stepantan
Nonionic	
Alkylphenol ethoxylates	Polystep F
Alkanolamides	Ninol
Fatty alcohol ethoxylates	Kessco
Fatty acid esters of glycerol and glycol	Kessco, Bio-Soft

STEPAN COMPANY (*Continued*)

Chemical name	Trade name
Zwitterionics and Amphoterics	
Alkyl amine oxide	Ammonyx
Alkyl amidopropyl amine oxides	Ammonyx
Alkyl amidopropyl betaines	Amphosol
Cationic	
Alkyl dimethyl benzyl ammonium chloride	Ammonyx
Alkyl trimethyl ammonium chloride	Ammonyx

TH GOLDSCHMIDT AG

Chemical name	Trade name
Nonionic	
Fatty acid ethoxylates	Tegin
Fatty alcohol ethoxylates	Teginacid, Tegotens
Fatty acid esters of ethylene glycol	Tegin
Fatty acid glycerol esters	Tegin
Amides	Tego-amide
Sucrose/glucose esters and derivatives	Tego Care, Tegosoft
Silicone surfactants	Abil
Zwitterionic	
Alkyl amino propyl betaines	Tego betaine
Alkyl amidopropyl amine oxide	Tegotens

UNIQEMA

Chemical name	Trade name
Anionic	
Alcohol ether phosphate esters	Monalube, Atlas, Monatrope, Atphos
Alkyl sulfosuccinates	Monawet, Monamate
Alkyl phenol ether sulfates	Atsurf
Nonionic	
Fatty alcohol ethoxylates	Arlasolve, Brij
Alkylphenol ethoxylates	Atlas, Atlox
Fatty acid ethoxylates	Alatone, Cirrasol, Renex
Fatty acid esters and sorbitan	Span, Atlox
Fatty acid esters of ethoxylated sorbitan	Tween, Atlox, Atsurf
Fatty acid esters of glycerol	Arlacel
Fatty acid ester ethoxylated glycols	Monalube
Fatty amine ethoxylates	Atlas
Alkanolamides	Monalube, Monamid
Betaines, Amphoterics	
Alkyl ampho acetates	Monateric
Alkyl ampho propionate	Monateric
Alkyl amido propyl betaine	Monateric
Alkyl amidazoline	Monoterge, Monazoline
Alkyl iminodipropionates	Monateric
Cationic	
Alkyl quaternary ammonium salts	Atlas
Alkyl morpholinium ammonium salts	Atlas
Polyoxyalkylene ammonium salts	Atlas

Index